Circuit Analysis Demystified

Demystified Series

Accounting Demystified
Advanced Calculus Demystified
Advanced Physics Demystified
Advanced Statistics Demystified
Algebra Demystified
Alternative Energy Demystified
Anatomy Demystified
asp.net 2.0 Demystified
Astronomy Demystified
Audio Demystified
Biology Demystified
Biotechnology Demystified
Business Calculus Demystified
Business Math Demystified
Business Statistics Demystified
C++ Demystified
Calculus Demystified
Chemistry Demystified
Circuit Analysis Demystified
College Algebra Demystified
Corporate Finance Demystified
Databases Demystified
Data Structures Demystified
Differential Equations Demystified
Digital Electronics Demystified
Earth Science Demystified
Electricity Demystified
Electronics Demystified
Engineering Statistics Demystified
Environmental Science Demystified
Everyday Math Demystified
Fertility Demystified
Financial Planning Demystified
Forensics Demystified
French Demystified
Genetics Demystified
Geometry Demystified
German Demystified
Home Networking Demystified
Investing Demystified
Italian Demystified
Java Demystified
JavaScript Demystified
Lean Six Sigma Demystified
Linear Algebra Demystified
Macroeconomics Demystified
Management Accounting Demystified
Math Proofs Demystified
Math Word Problems Demystified
MATLAB® Demystified
Medical Billing and Coding Demystified
Medical Terminology Demystified
Meteorology Demystified
Microbiology Demystified
Microeconomics Demystified
Nanotechnology Demystified
Nurse Management Demystified
OOP Demystified
Options Demystified
Organic Chemistry Demystified
Personal Computing Demystified
Pharmacology Demystified
Physics Demystified
Physiology Demystified
Pre-Algebra Demystified
Precalculus Demystified
Probability Demystified
Project Management Demystified
Psychology Demystified
Quality Management Demystified
Quantum Mechanics Demystified
Real Estate Math Demystified
Relativity Demystified
Robotics Demystified
Sales Management Demystified
Signals and Systems Demystified
Six Sigma Demystified
Spanish Demystified
sql Demystified
Statics and Dynamics Demystified
Statistics Demystified
Technical Analysis Demystified
Technical Math Demystified
Trigonometry Demystified
uml Demystified
Visual Basic 2005 Demystified
Visual C# 2005 Demystified
Vitamins and Minerals Demystified
xml Demystified

Circuit Analysis Demystified

David McMahon

New York Chicago San Francisco Lisbon London Madrid
Mexico City Milan New Delhi San Juan Seoul
Singapore Sydney Toronto

The McGraw-Hill Companies

Cataloging-in-Publication Data is on file with the Library of Congress.

McGraw-Hill books are available at special quantity discounts to use as premiums and sales promotions, or for use in corporate training programs. For more information, please write to the Director of Special Sales, Professional Publishing, McGraw-Hill, Two Penn Plaza, New York, NY 10121-2298. Or contact your local bookstore.

1 2 3 4 5 6 7 8 9 0 FGR/FGR 0 1 2 1 0 9 8 7

ISBN 978-0-07-148898-3
MHID 0-07-148898-7

This book is printed on acid-free paper.

Sponsoring Editor
Judy Bass

Acquisitions Coordinator
Rebecca Behrens

Editorial Supervisor
David E. Fogarty

Project Manager
Sandhya Joshi

Copy Editor
Nancy S. Wachter

Proofreader
Raj Kumar Singh

Indexer
Slavka Zlatkova

Production Supervisor
Pamela A. Pelton

Composition
Aptara, Inc.

Technical Editor
Rayjan Wilson

Cover Illustration
Lance Lekander

Art Director, Cover
Jeff Weeks

ABOUT THE AUTHOR

David McMahon is a physicist and researcher at a national laboratory. He is the author of *Linear Algebra Demystified, Quantum Mechanics Demystified, Relativity Demystified, Signals and Systems Demystified, Statics and Dynamics Demystified, and MATLAB® Demystified.*

CONTENTS

PREFACE

Circuit analysis is one of the most important courses in electrical engineering, where students learn the basics of the field for the first time. Unfortunately it is also one of the most difficult courses that students face when attempting to learn electrical engineering. At most universities it serves as a "weed out" course, where students not "cut out" for electrical engineering are shown the exit. A friend once referred to the course as "circuit paralysis" because he claimed to freeze up during the exams.

The purpose of this book is to make learning circuit analysis easier. It can function as a supplement to just about any electric circuits book and it will serve as a tutorial for just about any circuit analysis class. If you are having trouble with electrical engineering because the books are too difficult or the professor is too hard to understand, this text will help you.

This book explains concepts in a clear, matter-of-fact style and then uses solved examples to illustrate how each concept is applied. Quizzes at the end of each chapter include questions similar to the questions solved in the book, allowing you to practice what you have learned. The answer to each quiz question is provided at the end of the book. In addition, a final exam allows you to test your overall knowledge.

This book is designed to help students taking a one-year circuit analysis course or professionals looking for a review. The first 10 chapters cover topics typically discussed in a first-semester circuit analysis course, such as voltage and current theorems, Thevenin's and Norton's theorems, op amp circuits, capacitance and inductance, and phasor analysis of circuits.

The remaining chapters cover more advanced topics typically left to a second-semester course. These include Laplace transforms, filters, Bode plots, and characterization of circuit stability.

If you use this book for self-study or as a supplement in your class you will find it much easier to master circuit analysis.

ACKNOWLEDGMENTS

I would like to thank Rayjan Wilson for his thorough and thoughtful review of the manuscript. His insightful comments and detailed review were vital to making this book a success.

CHAPTER 1

An Introduction to Circuit Analysis

An *electric circuit* is an arrangement into a network of several connected *electric components*. The components that we will be concerned with are *two-terminal components*. This means that each component has two connection points or terminals that can be used to connect it with other components in the circuit. Each type of component will have its own symbol. This is illustrated in Fig. 1-1, where we indicate the terminals with two rounded ends or dots and use an empty box to represent a generic electric component.

There are several electric components but in this book our primary focus will be on resistors, capacitors, inductors, and operational amplifiers. At this point, we won't worry about what these components are. We will investigate each one in detail later in the book as the necessary theory is developed. In this chapter we will lay down a few fundamentals. We begin by defining *circuit analysis*.

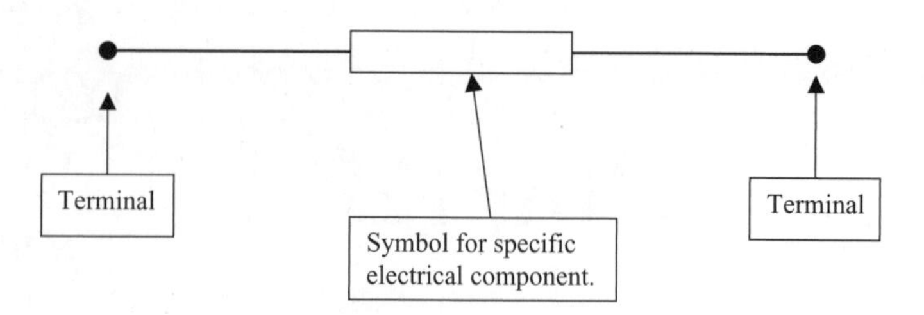

Fig. 1-1 A diagram of a generic two-terminal electric component.

What Is Circuit Analysis?

The main task of circuit analysis is to analyze the behavior of an electric circuit to see how it responds to a given input. The input could be a *voltage* or a *current,* or maybe some combination of voltages and currents. As you might imagine, electric components can be connected in many different ways. When analyzing a circuit, we may need to find the voltage across some component or the current through another component for the given input. Or we may need to find the voltage across a pair of output terminals connected to the circuit.

So, in a nutshell, when we do circuit analysis we want to find out how the unique circuit we are given responds to a particular input. The response of the circuit is the *output.* This concept is illustrated in Fig. 1-2.

To begin our study of circuit analysis, we will need to define some basic quantities like current and voltage more precisely.

Electric Current

Electric charge is a fundamental property of subatomic particles. The amount of electric charge that a particle carries determines how it will interact with

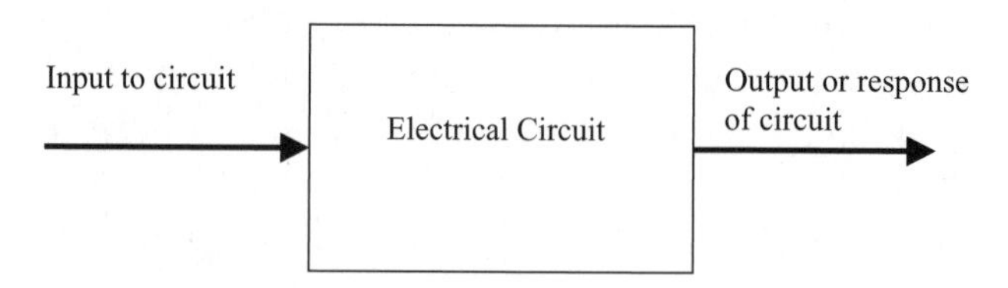

Fig. 1-2 The task of circuit analysis is to find out what the output or response of an electric circuit is to a given input, which may be a voltage or current.

electric and magnetic fields. In the SI system, which we will use exclusively in this book, the unit of charge is the *coulomb*. The symbol for a coulomb is *C*. An electron carries an electric charge given by

$$\text{charge of single electron} = 1.6 \times 10^{-19}\ \text{C} \tag{1.1}$$

The electric charge in an element or region can vary with time. We denote electric charge by $q(t)$, where the t denotes that charge can be a function of time.

The flow of charge or motion of charged particles is called *electric current.* We denote electric current by the symbol $i(t)$, where the t denotes that current can be a function of time. The SI unit for current is the *ampere* or *amp,* indicated by the symbol A. One amp is equal to the flow of one coulomb per second

$$1\ \text{A} = 1\ \text{C/s} \tag{1.2}$$

Current is formally defined as the rate of change of charge with time. That is, it is given by the derivative

$$i(t) = \frac{dq}{dt}\ \text{(amperes)} \tag{1.3}$$

EXAMPLE 1-1
The charge in a wire is known to be $q(t) = 3t^2 - 6$ C. Find the current.

SOLUTION
Using (1.3), we have

$$i(t) = \frac{dq}{dt} = \frac{d}{dt}(3t^2 - 6) = 6t\ \text{A}$$

EXAMPLE 1-2
Find the current that corresponds to each of the following functions of charge:

(a) $q(t) = 10 \cos 170\pi t$ mC
(b) $q(t) = e^{-2t} \sin t\ \mu$C
(c) $q(t) = 4e^{-t} + 3e^{5t}$ C

SOLUTION
In each case, we apply (1.3) paying special attention to the units. In (a), we have $q(t) = 10 \cos 170\pi t$ mC. Since the charge is measured in *millicoulombs*

or 10^{-3} C, the current will be given in *milliamps,* which is 10^{-3} A. Hence

$$i(t) = \frac{dq}{dt} = \frac{d}{dt}(10\cos 170\pi t) = -1700\pi \sin 170\pi t \text{ mA}$$

In (b), notice that the charge is expressed in terms of *microcoulombs*. A microcoulomb is 10^{-6} C, and the current will be expressed in microamps. Using the product rule for derivatives which states

$$(fg)' = f'g + g'f$$

We find that the current is

$$\begin{aligned} i(t) &= \frac{dq}{dt} = \frac{d}{dt}(e^{-2t}\sin t) \\ &= \frac{d}{dt}(e^{-2t})\sin t + e^{-2t}\frac{d}{dt}(\sin t) \\ &= -2e^{-2t}\sin t + e^{-2t}\cos t \\ &= e^{-2t}(-2\sin t + \cos t)\ \mu\text{A} \end{aligned}$$

Finally, in (c), the charge is given in coulombs, and therefore, the current will be given in amps. We have

$$i(t) = \frac{dq}{dt} = \frac{d}{dt}(4e^{-t} + 3e^{5t}) = -4e^{-t} + 15e^{5t} \text{ A}$$

Looking at (1.3), it should be apparent that, given the current flowing past some point P, we can integrate to find the total charge that has passed through the point as a function of time. Specifically, let's assume we seek the total charge that passes in a certain interval that we define as $a \le t \le b$. Then given $i(t)$, the charge q is given by

$$q = \int_a^b i(t)\,dt \tag{1.4}$$

EXAMPLE 1-3

The current flowing through a circuit element is given by $i(t) = 8t + 3$ mA. How much charge passed through the element between $t = 0$ and $t = 2$ s?

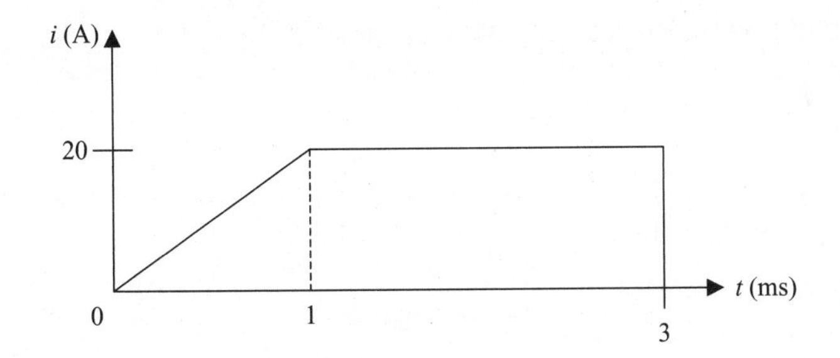

Fig. 1-3 A plot of the current flowing past some point in a circuit.

SOLUTION
We can find the total charge that passed through the element by using (1.4). We have

$$q = \int_a^b i(t)\, dt = \int_0^2 (8t + 3)\, dt = 8 \int_0^2 t\, dt + 3 \int_0^2 dt$$
$$= 4t^2 \Big|_0^2 + 3t \Big|_0^2 = (16 + 6) \text{ mC} = 22 \text{ mC}$$

EXAMPLE 1-4
The current flowing past some point is shown in Fig. 1-3. Find the total charge that passes through the point.

SOLUTION
First, notice that time is given in *milliseconds* and current is given in *amps*. Looking at the definition of the amp (1.2), we could write the coulomb as

$$1 \text{ C} = 1 \text{ A-s}$$

Looking at the definition (1.4), the integrand is the product of current and time. In this example, as we stated above, current is given in amps and time is given in ms $= 1 \times 10^{-3}$ s. Therefore the final answer should be expressed as

$$(1 \text{ A})(1 \text{ ms}) = 1 \times 10^{-3} \text{ A-s} = 10^{-3} \text{ C} = 1 \text{ mC}$$

Now let's look at the plot. It is divided into two regions characterized by a different range of time. We can find the total charge that flows past the point by finding the total charge that flows in each range and then adding the two charges together. We call the total charge that flows past the point for $0 \le t \le 1$ q_1 and we denote the total charge that flows past the point for $1 \le t \le 3$ q_2. Once we calculate these quantities, our answer will be

$$q = q_1 + q_2 \tag{1.5}$$

The first region is defined for $0 \leq t \leq 1$ where the current takes the form of a straight line with a slope

$$i(t) = at + b \text{ A}$$

where a and b are constants. We know the value of the current at two points

$$i(0) = 0 \text{ A}, \quad i(1) = 20 \text{ A}$$

First, using $i(0) = 0$ together with $i(t) = at + b$ tells us that $b = 0$, so we know the current must assume the form $i(t) = at$ A. Second, $i(1) = 20$ A allows us to determine the value of the constant a, from which we find that $a = 20$. Therefore

$$q_1 = \int_0^1 i(t)\, dt = 20 \int_0^1 t\, dt = \frac{20}{2} t^2 \Big|_0^1 = 10 \text{ mC} \tag{1.6}$$

As an aside, what are the units of a? If $i(t) = at$ A then the product at must be given in amperes. Remembering that t is given in milliseconds

$$[at] = [a]\,[\text{ms}] = \text{A} = \text{C/s}$$

$$\Rightarrow [a] = \frac{\text{C}}{\text{ms-s}}$$

There are 10^{-3} s in a millisecond, therefore

$$[a] = \frac{\text{C}}{\text{ms-s}} = \left(\frac{\text{C}}{\text{ms-s}}\right)\left(\frac{10^{-3}\text{ s}}{\text{ms}}\right) = \frac{10^{-3}\text{ C}}{(\text{ms})^2} = \frac{\text{mC}}{(\text{ms})^2}$$

Notice how this is consistent with (1.6), where we integrate over 0 to 1 ms, and we have a factor of time squared that cancels the time squared in the denominator of the units used for the constant a, leaving millicoulombs in the final result.

Let's finish the problem by examining the region defined by $1 \text{ ms} \leq t \leq 3$ ms. In this region, the current is a constant given by $i(t) = 20$ A. The total charge that passes is

$$q_2 = \int_1^3 i(t)\, dt = 20 \int_1^3 dt = 20t\big|_1^3 = 40 \text{ mC}$$

In conclusion, using (1.5) the total charge that passes the point is

$$q = 10 \text{ mC} + 40 \text{ mC} = 50 \text{ mC}$$

The next example will be a little bit painful, but it will help us review some calculus techniques that come up frequently in electrical engineering.

EXAMPLE 1-5

The current flowing through a circuit element is given by $i(t) = e^{-3t} 16 \sin 2t$ mA. How much charge passed through the element between $t = 0$ and $t = 3$ s?

SOLUTION

We can find the total charge that passed through the element by using (1.4). We have

$$q = \int_a^b i(t)\, dt = \int_0^3 e^{-3t}\,(16 \sin 2t)\, dt$$
$$= 16 \int_0^3 e^{-3t} \sin 2t\, dt \text{ mC}$$

We can do this problem using integration by parts. The integration-by-parts formula is

$$\int f(t) \frac{dg}{dt}\, dt = f(t)g(t) - \int g(t) \frac{df}{dt}\, dt \tag{1.7}$$

Looking at the integral in our problem, we let

$$f(t) = e^{-3t} \Rightarrow \frac{df}{dt} = -3e^{-3t}$$

This means that

$$\frac{dg}{dt} = \sin 2t$$

Using elementary integration we find that

$$g(t) = -\frac{1}{2} \cos 2t$$

So using (1.7), we have

$$16\int_0^3 e^{-3t}\sin 2t dt = 16\left(-\frac{1}{2}e^{-3t}\cos 2t\Big|_0^3 - \frac{3}{2}\int_0^3 e^{-3t}\cos 2t\,dt\right)$$

$$= -8e^{-3t}\cos 2t\Big|_0^3 - 24\int_0^3 e^{-3t}\cos 2t\,dt$$

Now we have to apply integration by parts again on the second term. Using the same procedure where we make the identification

$$\frac{dg}{dt} = \cos 2t$$

We find that

$$\int_0^3 e^{-3t}\cos 2t\,dt = \frac{1}{2}e^{-3t}\sin 2t\Big|_0^3 + \frac{3}{2}\int_0^3 e^{-3t}\sin 2t\,dt$$

Hence

$$16\int_0^3 e^{-3t}\sin 2t\,dt = -8e^{-3t}\cos 2t\Big|_0^3 - 24\int_0^3 e^{-3t}\cos 2t\,dt$$

$$= -8e^{-3t}\cos 2t\Big|_0^3 - 12e^{-3t}\sin 2t\Big|_0^3 - 36\int_0^3 e^{-3t}\sin 2t\,dt$$

Now we add $36\int_0^3 e^{-3t}\sin 2t\,dt$ to both sides. This gives the result

$$52\int_0^3 e^{-3t}\sin 2t\,dt = -8e^{-3t}\cos 2t\Big|_0^3 - 12e^{-3t}\sin 2t\Big|_0^3$$

The right-hand side evaluates to

$$-8e^{-3t}\cos 2t\Big|_0^3 - 12e^{-3t}\sin 2t\Big|_0^3 = -8e^{-9}\cos 6 + 8 - 12e^{-9}\sin 6 \approx 8$$

Therefore

$$\int_0^3 e^{-3t}\sin 2t\,dt = \frac{1}{52}\left(-8e^{-3t}\cos 2t\Big|_0^3 - 12e^{-3t}\sin 2t\Big|_0^3\right) \approx \frac{8}{52} = 0.154$$

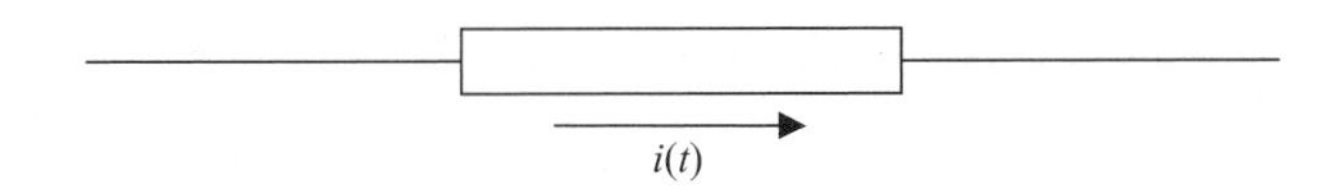

Fig. 1-4 We indicate current in a circuit by drawing an arrow that points in the direction of current flow.

So the total charge is

$$q = 16\int_0^3 e^{-3t}\sin 2t\,dt \text{ mC} = (16)(0.154)\text{ mC} = 2.461\text{ mC}$$

Current Arrows

When drawing an electric circuit, the *direction* of the current is indicated by an arrow. For example, in Fig. 1-4 we illustrate a current flowing to the right through some circuit element.

The flow of current can be defined by the flow of positive charge or the flow of negative charge. Even though we think of current physically as the flow of electrons through a wire, for instance, by convention in electrical engineering we measure current as the rate of flow of *positive charge*. Therefore

- A current arrow in a circuit diagram indicates the direction of flow of positive charge.
- A positive charge flow in one direction is equivalent to a negative charge flow in the opposite direction.

For example, consider the current shown flowing to the right in Fig. 1-4. Finding that $i(t) > 0$ when we do our calculations means that positive charges are flowing in the direction shown by the arrow. That is,

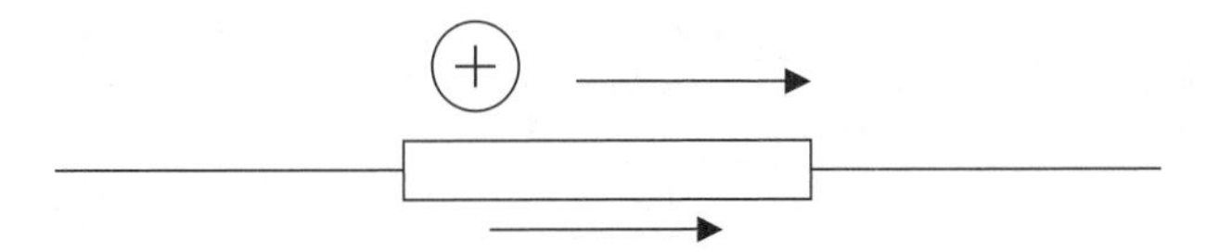

$i(t) > 0 \Rightarrow$ Positive charges flowing in direction of arrow

Now suppose that when we do the calculations, we instead find that $i(t) < 0$. This means that the positive charges are actually flowing in the direction opposite to that indicated by the arrow. In this case we have the following situation:

If $i(t) > 0 \Rightarrow$ Positive charges flowing in direction of arrow

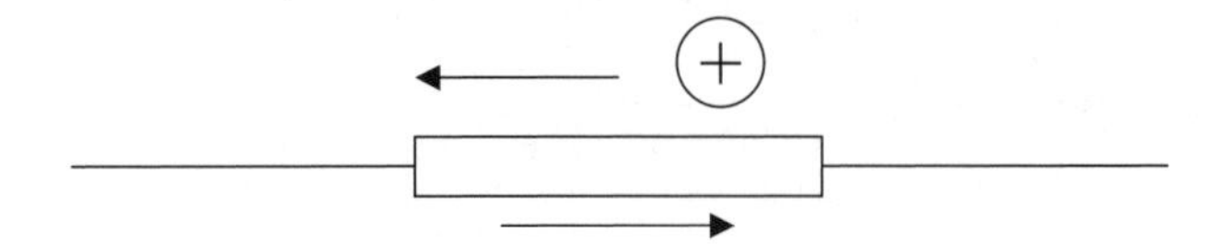

$i(t) < 0 \Rightarrow$ Positive charges are flowing in direction opposite to the arrow

Since the current in this case is calculated to be negative, this is equivalent to a positive current flowing in the opposite direction. That is, we reverse the direction of the arrow to take $i(t)$ to be positive.

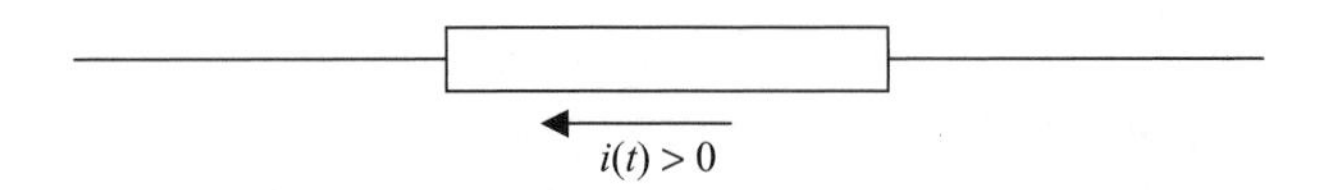

Let's focus on this point for a minute by looking at some examples. This means that a flow of +5 C/s to the right is the same as −5 C/s flowing to the left. It also means that 7 A of negative charge flowing to the left is equivalent to 7 A of positive charge flowing to the right.

EXAMPLE 1-6

At a certain point P in a wire, 32 C/s flow to the right, while 8 C/s of negative charge flow to the left. What is the net current in the wire?

SOLUTION

By convention we define current as the rate of flow of positive charge. The current that flows to the right in the wire is

$$i_R(t) = +32 \text{ A}$$

The current flowing to the left is negative charge

$$i_L(t) = -8 \text{ A}$$

Now 8 A of negative charge flowing to the left is equivalent to 8 A of positive charge flowing to the right. So the net current is

$$i(t) = i_R(t) - i_L(t) = 32 - (-8) = 40 \text{ A}$$

Let's combine the idea of positive charge flow with the representation of current in a circuit diagram with a little arrow, as in Fig. 1-4. With the convention that the arrow points in the direction of positive charge flow

- If the value of the current satisfies $i(t) > 0$, then positive charges are flowing *in the direction* that the arrow points.
- If the value of the current satisfies $i(t) < 0$, then the flow of positive charge is *in the direction opposite* to that indicated by the arrow.

Refer to Fig. 1-4 again. If we are told that $i(t) = 6$ A, then this means that 6 A of positive charge are flowing to the right in the circuit. On the other hand, if we are told that $i(t) = -3$ A, then this means that 3 A of positive charge are flowing to the left in the circuit. The negative sign means that the flow of positive charge is in the direction opposite to that indicated by the arrow. Hence, while the current arrow is to the right, since $i(t) = -3$ A, which is less than zero, the positive charges are flowing to the left:

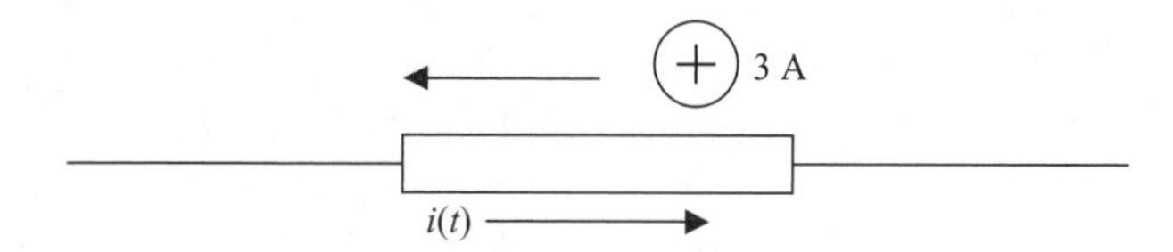

$i(t) < 0 \Rightarrow$ Positive charges are flowing in direction opposite to the arrow

Voltage

The next part of the basic foundation we need to add to our toolkit for studying electric circuits is the concept of *voltage*. In short, voltage is the electric version of *potential energy,* which is energy that has the potential to do work. The first example of potential energy that a student encounters is usually the potential energy of a mass m in a gravitational field g. If the mass m is at a height h with respect to some *reference point,* then the potential energy is

$$U = mgh$$

The gravitational potential energy has meaning only when it is thought of as a potential difference between two heights. If the mass falls from the upper height to the lower height, it gains kinetic energy. The mass obtains the energy from the potential U. Recall from your studies of elementary physics that when using SI units we measure energy in *joules*, which are indicated by the symbol J.

Voltage is analogous to potential energy, and it is often referred to as the *potential difference* between two points A and B in a circuit. The units of voltage are

$$\begin{aligned} &1 \text{ volt} = 1 \text{ joule/coulomb} \Rightarrow \\ &1 \text{ V} = 1 \text{ J/C} \end{aligned} \tag{1.8}$$

In circuit analysis we usually indicate voltage as a function of time by writing $v(t)$. The voltage between points A and B in a circuit is the amount of energy required to move a charge of 1 C from A to B. Voltage can be positive or negative. When the voltage is positive, i.e., $v(t) > 0$, we say that the path A–B is a *voltage drop*. When a positive charge passes through a voltage drop, the charge *gains* energy. This is because, if $v(t) > 0$, the point A is at a higher potential than the point B, in the same way that a point 100 m above the surface of the earth is at a higher potential than a point at sea level, since $U = mgh$ for a gravitational field.

On the other hand, suppose that the voltage between two points A and B in a circuit is negative. In this case, we say that the path A–B is a *voltage rise*. To move a positive charge from A to B when the path is a voltage rise, we have to *supply* energy. This is analogous to the energy you have to supply to lift a 50 lb weight from the ground to a spot on the shelf 5 ft higher.

Voltage is formally defined as

$$v = \frac{dw}{dq} \text{ V} \tag{1.9}$$

where w is the work required to move the charge w across the potential difference.

To find the energy acquired by a charge, we examine the units of voltage, which are given as joules per coulomb which is energy per charge. Therefore to find the energy that a charge gains or loses when passing through a potential difference, we multiply the charge carried by the voltage

$$E = qV \tag{1.10}$$

EXAMPLE 1-7

A 2 C charge and a −7 C charge pass through a potential difference of +3 V and a potential difference of −2 V. Find the energy gained or lost by each charge.

SOLUTION

We apply (1.10). When the 2 C charge passes through the potential difference of +3 V

$$E = qV = (2\text{ C})(3\text{ V}) = (2\text{ C})(3\text{ J/C}) = 6\text{ J}$$

This means that 6 J of energy had to be added to the system to move the charge through the potential difference. When the charge passes through the potential difference of −2 V

$$E = qV = (2\text{ C})(-2\text{ V}) = (2\text{ C})(-2\text{ J/C}) = -4\text{ J}$$

Since the energy is negative, the charge acquired or gained 4 J of energy when passing through the potential difference. Now let's consider the −7 C charge. When this charge passes through the first potential difference

$$E = qV = (-7\text{ C})(3\text{ V}) = (-7\text{ C})(3\text{ J/C}) = -21\text{ J}$$

This charge acquired 21 J of energy moving through the 7 V potential. In the second case

$$E = qV = (-7\text{ C})(-2\text{ V}) = (-7\text{ C})(-2\text{ J/C}) = 14\text{ J}$$

The energy is positive, indicating that the charge lost energy moving through the potential difference.

Time Varying Voltage and Voltage Sources

We are all used to the terms *DC* and *AC* and have seen constant voltage sources like 12 V for a battery. Although we may be used to 9 and 12 V batteries, in many situations the voltage in a circuit will vary with time. We have already indicated this by writing voltage as a time-dependent function $v(t)$. Of particular interest are voltages that oscillate *sinusoidally*. For example, in the United States, the voltage in a household outlet oscillates between +170 and −170 V according to

$$v(t) = 170 \sin 377t \tag{1.11}$$

In general, a sinusoidal function can be written as

$$f(t) = A \sin \omega t \tag{1.12}$$

We call A the *amplitude* of the sine wave. The units of the amplitude depend on the type of wave that is oscillating. In (1.11), the amplitude is $A = 170$ V. In short, the amplitude is the maximum height that the function attains above the origin.

The *angular frequency* of the sine wave is given by ω. This is related to the *frequency*, which is denoted by ν using the relation

$$\omega = 2\pi\nu \tag{1.13}$$

Angular frequency is measured in radians per second. The frequency ν tells us the number of cycles per second in the wave. A cycle is a complete repetition of the waveform; therefore, the number of cycles per second is the number of times the waveform repeats in one second. We can abbreviate cycles per second by writing *cps* and note that a cycle per second is a *hertz*

$$1 \text{ cps} = 1 \text{ Hz} \tag{1.14}$$

For a U.S. household voltage in (1.11), the angular frequency is $\omega = 377$ rad/s and the number of cycles per second is

$$\nu = \frac{377}{2\pi} = 60 \text{ cps} = 60 \text{ Hz} \tag{1.15}$$

The amplitude and cycle for (1.11) are shown in Fig. 1-5.

In a circuit, we can supply energy with a *voltage source.* As far as the circuit is concerned, the voltage source can be a "black box." The internal details or

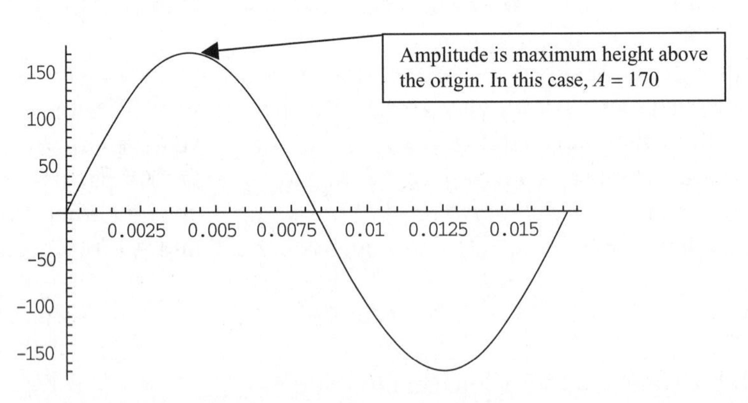

Fig. 1-5 A plot of $v(t) = 170 \sin 377t$. The plot shows exactly one cycle. To show one complete cycle, we plot from $t = 0$ to $t = 1/60$ s.

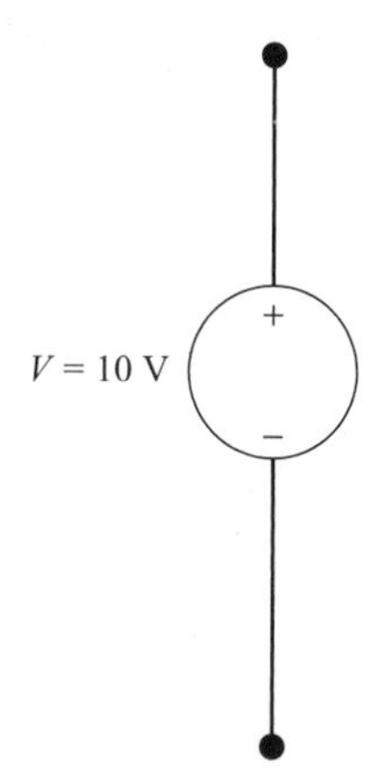

Fig. 1-6 A 10 V voltage source.

construction of the voltage source do not concern us; it can be any electric element that maintains a specific voltage across its terminals. For the purposes of circuit analysis we want to see what happens when the voltage serves as an input to excite the circuit. Then we do analysis to see what the response of the circuit will be.

We indicate a voltage source in a circuit by drawing a circle and show the positive and negative reference points for the voltage. In Fig. 1-6, we indicate a voltage source such that moving down along the element gives a voltage of +10 V while going up along the element would give a voltage of −10 V.

Besides keeping the direction of the voltage straight, the key concept to understand is that a voltage source maintains the voltage indicated at all times *no matter what other elements* are connected to it. However, the behavior of the voltage source is not completely independent from the rest of the circuit. The other elements in the circuit determine the current that flows through it.

In Fig. 1-7, we show a circuit consisting of some voltages. To write down the value of the voltages, we go around the circuit in a clockwise direction.

Starting at the 10 V voltage source, if moving clockwise we are moving from − to + across the voltage source, so we pick up −10 V. Going around up to the 3 V source, we have

$$-10\text{ V} - 6\text{ V} + 4\text{ V}$$

Now when we get to the 3 V source, we are moving in the opposite way to what we did at the 10 V source; that is, we are moving from + to − and so we

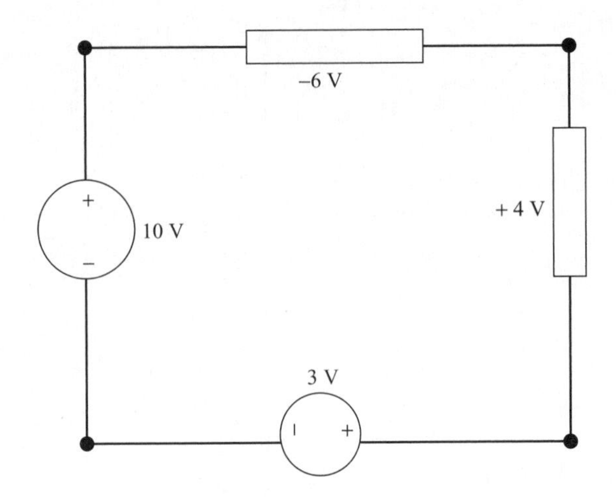

Fig. 1-7 An illustration of how to add up voltages in a circuit.

add +3 V, giving the complete path around the circuit

$$-10\ \text{V} - 6\ \text{V} + 4\ \text{V} + 3\ \text{V}$$

Dependent Voltage Sources

We can also have voltage sources whose values are *dependent* on some other element in the circuit. A dependent source is indicated with a diamond shape. For example, if there is some current $i(t)$ in the circuit, a voltage source that varies with $i(t)$ as $v(t) = ri(t)$, where r is a constant is illustrated by the diamond shown in Fig. 1-8.

Current Sources

We can also "excite" a circuit by supplying a current from an external source. In the same way that a voltage source can be thought of as a black box, the internal construction of a current source is not of any concern in circuit analysis. In a real circuit, a current source may be a transistor circuit that supplies current to some other circuit that is being analyzed. But we don't care what the internal construction is—we only care about what current $i(t)$ the current source supplies. Simply put, a current source is a circuit element that *always* has a specified current flowing through it. A current source behaves in an inverse manner to a

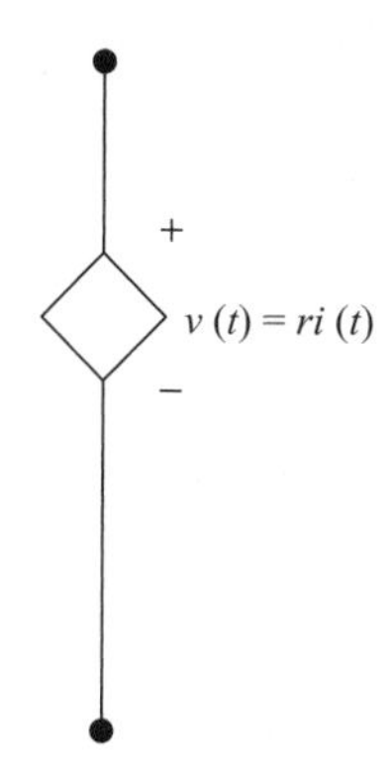

Fig. 1-8 A dependent voltage source.

voltage source. While a voltage source operates at a fixed voltage and the current flowing through it is determined by the other components in the circuit, a current source always has a specified current flowing through it and the voltage across it is determined by what elements are connected to it in the circuit.

A current source is shown in a circuit diagram by drawing a circle that contains a current arrow in it. As usual, the arrow indicates the direction of flow of positive charge. An example is shown in Fig. 1-9.

It is also possible to have *dependent current sources*. These are indicated with a diamond shape containing an arrow indicating the direction of the current. The current that flows through the element can be dependent on some other quantity in the circuit. For example, the current can be dependent on some voltage $v(t)$ using the relation $i(t) = gv(t)$, where g is a constant. This is shown in Fig. 1-10.

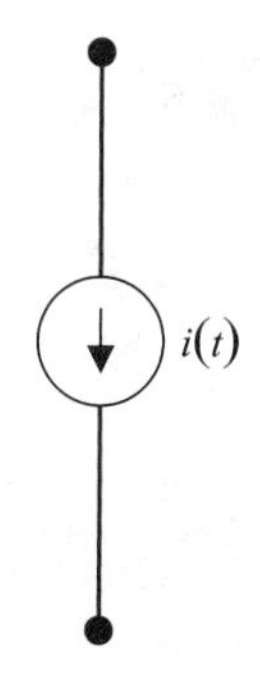

Fig. 1-9 A current source.

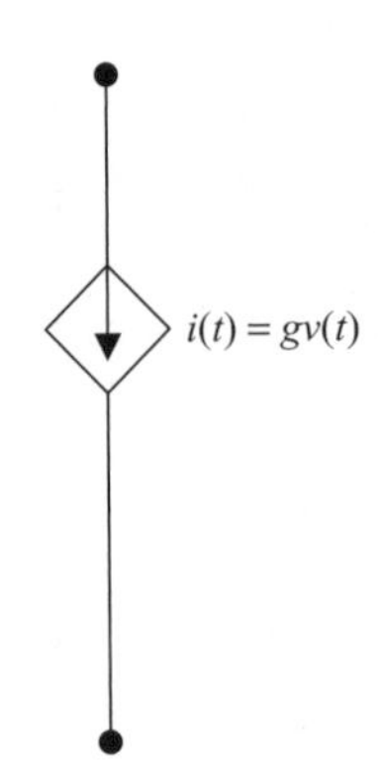

Fig. 1-10 A dependent current source.

Open and Short Circuits

We are nearly done with our tour of the foundational elements of circuit analysis. Next we consider two terms that are common in the English language, *open circuits* and *closed circuits*. Your common sense view of an open circuit is probably accurate. You can think of it as basically an open switch. In fact, as shown in Fig. 1-11, we indicate an open circuit by a drawing of an open switch in a circuit diagram.

As you might guess from the fact that an open circuit has a switch that is open, there is *no conducting path* through an open circuit. In other words the current through an open circuit is $i = 0$. However, an open circuit *can* have a voltage across it.

A *short circuit* has no voltage across it, so $v = 0$. However, a short circuit is a *perfectly conducting path*. We indicate a short circuit by drawing a straight line in a circuit diagram.

Fig. 1-11 An open circuit

Power

We conclude this introductory chapter with a look at *power*. The SI unit used for power is the *watt*, where

$$1\ \text{W} = 1\ \text{J/s} \tag{1.16}$$

We often write W to indicate watts. In electric circuit analysis, power is the product of voltage and current. Recall that the units of voltage are joules/coulomb and the units of current are coulombs/second, so if we form the product coulombs cancel giving joules/second or watts. If we denote power by $p(t)$, then

$$p(t) = v(t)\,i(t) \tag{1.17}$$

The power in a circuit element can be positive or negative, and this tells us whether or not the circuit element *absorbed power* or if it is a *power supply*. If the power in a circuit element is positive

$$p = vi > 0$$

then the element absorbs power. If the power is negative

$$p = vi < 0$$

then the element *delivers* power to the rest of the circuit. In other words it is a power supply.

When analyzing the power in a circuit, we examine the direction of the current arrow relative to the signs indicated for the voltage. If the current arrow points in the direction from the + to − signs along the voltage (i.e., along a voltage drop), then the power is positive. This is shown in Fig. 1-12.

Remember that, if the current in Fig. 1-12 is negative, the power will be negative as well. So if $v(t) = 5$ V and $i(t) = 3$ A, the power for the element in Fig. 1-12 is

$$p = (5\ \text{V})(3\ \text{A}) = 15\ \text{W}$$

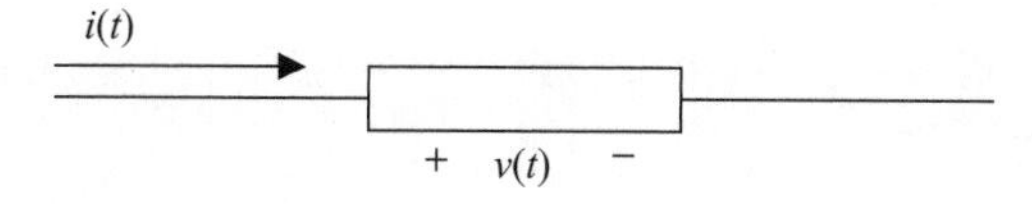

Fig. 1-12 The power is $p = (+v)(+i) = vi$.

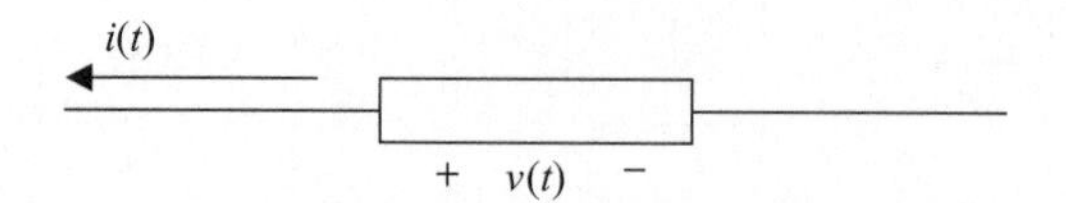

Fig. 1-13 If the current arrow points away from the positive point of a voltage, use $-i$ when doing power calculations.

Since the power is positive, the element *absorbs* power. On the other hand, suppose that $i(t) = -3$ A. Then

$$p = (5 \text{ V})(-3 \text{ A}) = -15 \text{ watts}$$

In this case, the power is negative and the element delivers power. The element is a power supply. Note that a given circuit element can be a power supply or absorb power at different times in the *same* circuit, since the voltages and currents may vary with time.

If the current arrow points in the opposite direction to the +/− terminals of the voltage source, we take the *negative* of the current when computing the power. This is shown in Fig. 1-13.

We repeat the calculations we did for the circuit element shown in Fig. 1-12. This time, looking at Fig. 1-13, we need to reverse the sign of the current. If $v(t) = 5$ V and $i(t) = 3$ A, the power for the element in Fig. 1-13 is

$$p = (5 \text{ V})(-3 \text{ A}) = -15 \text{ W}$$

Since the power is negative, the element delivers power. On the other hand, suppose that $i(t) = -3$ A. Then

$$p = (5 \text{ V})(-(-3\text{A})) = +15 \text{ W}$$

In other words, the circuit absorbs power.

EXAMPLE 1-8

Determine the power supplied or absorbed for each element in the circuit shown in Fig. 1-14.

SOLUTION

Starting on the left, we begin our analysis of the 10 V voltage source. A 2 A current is flowing *away* from the positive terminal of the voltage source. Therefore, the power is

$$p_1 = (11 \text{ V})(-2 \text{ A}) = -22 \text{ W}$$

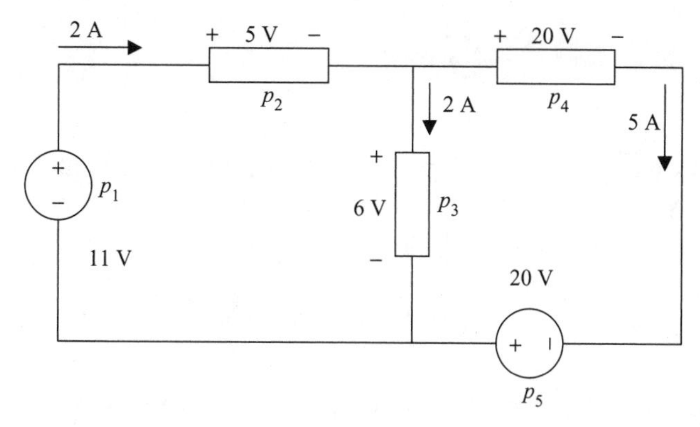

Fig. 1-14 The circuit analyzed in Example 1-8.

The power is negative, so the element delivers power. Moving to element 2, now the current points from the + to − terminal of the voltage. Therefore, we do not change the sign of the current. So in this case the power is

$$p_2 = (5\text{ V})(2\text{ A}) = 10\text{ W}$$

Since the power is positive, element 2 absorbs power.

Moving on to element 3, the current points from the + to − terminal of the voltage. The power is

$$p_3 = (6\text{ V})(2\text{ A}) = 12\text{ W}$$

Element 3 also absorbs power. The current flowing through element 4 is the 5 A current on the right side of the circuit diagram. This current also flows from positive to negative as indicated by the voltage, so the power is

$$p_4 = (20\text{ V})(5\text{ A}) = 100\text{ W}$$

Finally, we arrive at element 5. In this case, although the magnitudes of the current and voltage are the same, the current flows from the negative to the positive terminals of the voltage source, so the power is

$$p_5 = (20\text{ V})(-5\text{ A}) = -100\text{ W}$$

The power is negative; therefore, the element 5 is a power supply that delivers power to the circuit.

Conservation of Energy

If you add up the power calculated for each of the elements in Example 1-8, you will find that they sum to zero. This is a general principle that we can use in circuit analysis. The conservation of energy tells us that if we sum up all the power in an electric circuit, the total power is zero

$$\sum p_i = 0 \tag{1.18}$$

This principle can be used to find an unknown power in a circuit.

EXAMPLE 1-9

For the circuit shown in Fig. 1-15, find the power in element 3 using conservation of energy.

SOLUTION

Conservation of energy tells us that

$$\sum p_i = 0$$

For the circuit shown

$$\sum p_i = 100 - 20 + p_3 = 0$$

Moving all the relevant terms to the right side we find that

$$p_3 = -100 + 20 = -80 \text{ W}$$

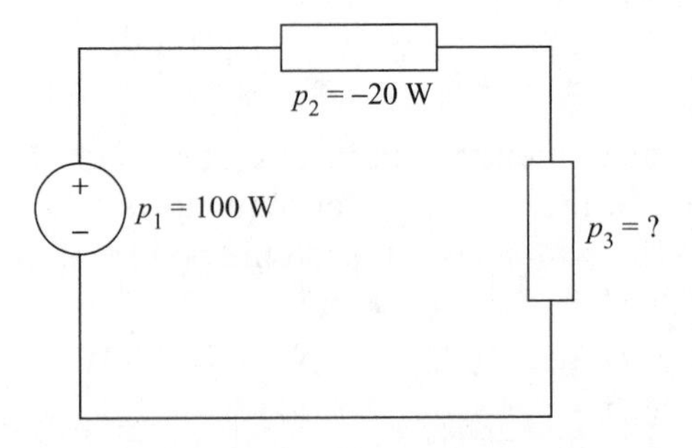

Fig. 1-15 We can find the power in the third element by applying conservation of energy.

Summary

In this chapter we have introduced some basic notions that form the foundation of circuit analysis. We have learned that current is the amount of charge that flows per second and that in electrical engineering, by convention, we indicate the direction of positive charge flow in a circuit by a current arrow. We have also learned about voltage and current sources and how to calculate power in a circuit. All of the examples in this chapter have used generic circuit elements. In the next chapter, we start to examine real electric circuits by considering our first circuit element, the *resistor.*

Quiz

1. You establish an observation point in a wire and find that $q(t) = 2t$ C. Find the current flowing past your observation point.
2. If $q(t) = 10e^{-2t} \cos 5t$ mC, what is the corresponding current? Plot the current as a function of time from 0 to 2 s.
3. If the current is $i(t) = 150 \sin 77t$, where current is given in amps, how much charge flows by between 0 and 5 s?
4. At a certain point P in a wire, 20 C of positive charge flow to the right while 8 C of negative charge flow to the left. What is the current flowing in the wire?
5. A charge $q = 7$ C passes through a potential difference of 8 V. How much energy does the charge acquire?

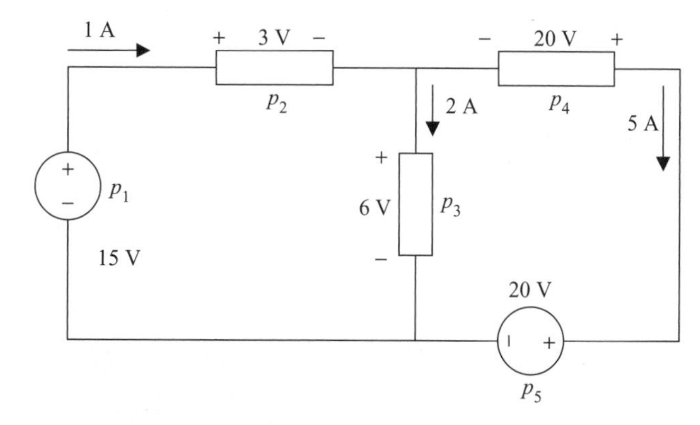

Fig. 1-16 Circuit diagram for Problem 8.

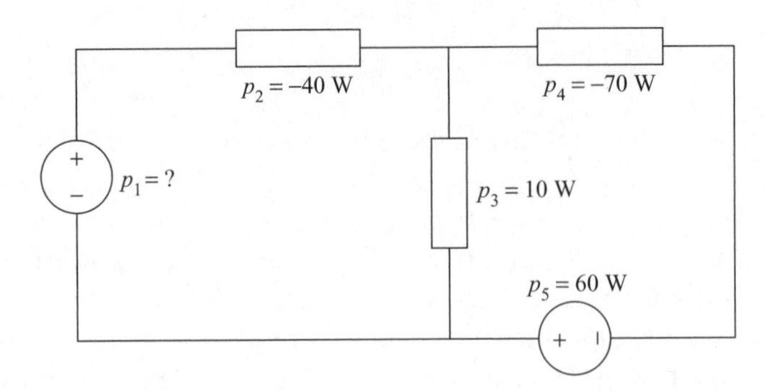

Fig. 1-17 Circuit diagram for Problem 10.

6. If the voltage in a circuit is given by $v(t) = 120 \cos 200\pi t$, what is are the amplitude and cycles per second?
7. In some circuit element the power is 20 W and the voltage is 10 V. How much current flows?
8. Find the power in each element shown in Fig. 1-16.
9. How does conservation of energy manifest itself in a circuit?
10. Find the missing power in Fig. 1-17.

CHAPTER 2

Kirchhoff's Laws and Resistance

In this chapter we will encounter two laws that are general enough to apply to any circuit. The first of these, *Kirchhoff's current law,* is a result of the conservation of charge and tells us that the sum of currents at a connection point in a circuit must vanish. The second law, which derives from the conservation of energy, is *Kirchhoff's voltage law*. This law tells us that the sum of voltages in a closed path in a circuit must vanish.

After describing these laws in more detail, we will consider the concept of *resistance* and meet our first real element in circuit analysis, the *resistor*.

Branches, Nodes, and Loops

In this section we lay out some definitions that will be important throughout the book. A *branch* is a single element or component in a circuit. If several elements in a circuit carry the same current, they can also be referred to as a branch.

A *node* is a connection point between two or more branches—which usually means a connection point between two or more elements in a circuit. We indicate the presence of a node in a circuit with a large dot. Physically, a connection point in a circuit is a point where two or more elements have been soldered together. Kirchhoff's current law applies to nodes.

A *loop* is a closed path in a circuit. Kirchhoff's voltage law applies to loops.

Kirchhoff's Current Law

As we indicated in the introduction, Kirchhoff's current law, which we will refer to from now on as KCL, is a consequence of the conservation of charge. This fundamental principle of physics tells us that, in a volume of space, charge cannot be created or destroyed. If charges are flowing through the region of interest, another way to express this principle is to say that the amount of charge entering the region is equal to the amount of charge leaving the region.

A node is a single point at which we can apply the conservation of charge. Charge cannot accumulate or be destroyed at a node in a circuit. Said another way, *the amount of charge entering a node must be equal to the amount of charge leaving the node*. We can express this fact mathematically by saying that the sum of all currents at a node must vanish. That is,

$$\sum i(t) = 0 \tag{2.1}$$

KCL is applied at each node in a circuit and holds in general. To reflect that the current flowing into a node added to the flow of current out of a node vanishes, we must assign positive and negative values to these currents. The choice is entirely arbitrary and is up to you, but whatever choice you make must be applied to every node in the circuit you are analyzing. In this book, we choose to apply the following convention:

- $+$ for currents entering a node
- $-$ for currents leaving a node

For example, consider Fig. 2-1, which shows the current i_1 entering the node and the current i_2 leaving the node.

The current i_1 enters the node; therefore, applying our convention we take it to be positive. On the other hand, the current i_2 is leaving the node, so we take

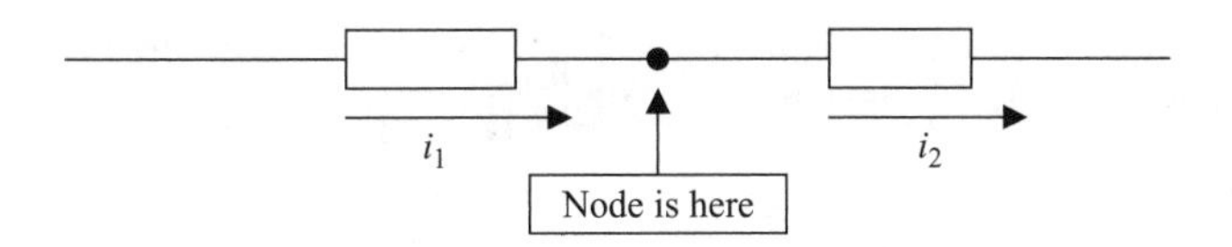

Fig. 2-1 A node representing the connection point between two circuit elements.

it to be negative. KCL at this node is then written as

$$\sum i = i_1 - i_2 = 0$$

Let's apply KCL to a more substantial example.

EXAMPLE 2-1
Consider the circuit shown in Fig. 2-2. If $i_1 = 3$ A, $i_3 = 5$ A, $i_4 = 6$ A, and $i_5 = 1$ A, find i_2.

SOLUTION
KCL tells us that the sum of the currents at the node shown in Fig. 2-2 must vanish. That is,

$$\sum i_n = 0$$

Taking + for currents entering the node and − for currents leaving the node, KCL gives us

$$i_1 - i_2 - i_3 + i_4 - i_5 = 0$$

Solving for i_2,

$$i_2 = i_1 - i_3 + i_4 - i_5 = 3 - 5 + 6 - 1 = 3 \text{ A}$$

EXAMPLE 2-2
Consider the node shown in Fig. 2-3. If $i_1 = 2$ A and $i_2 = 7$ A, find i_3.

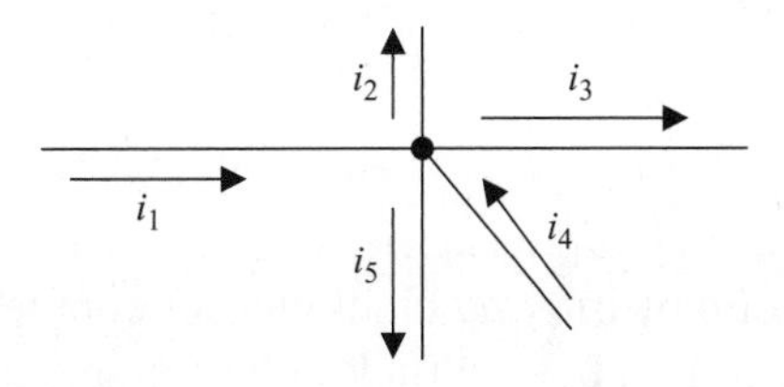

Fig. 2-2 Currents at a node used in Example 2-1.

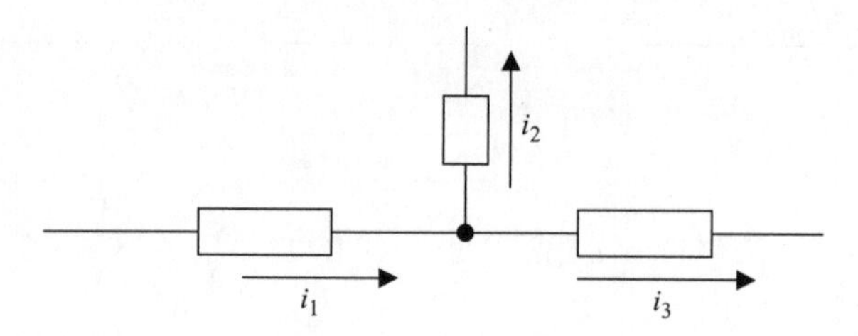

Fig. 2-3 Currents entering and leaving the node studied in Example 2-2.

SOLUTION

Again, KCL tells us that the currents entering the node added to the currents leaving the node must vanish. Taking + for currents entering the node and – for currents leaving the node, we have

$$i_1 - i_2 - i_3 = 0$$

Solving for the unknown current, we obtain

$$i_3 = i_1 - i_2 = 2 - 7 = -5 \text{ A}$$

In this case we obtain a *negative* answer. This tells us that the actual flow of current is in the direction opposite to that we chose for the arrow in Fig. 2-3. That is, i_3 must actually be *entering* the node given the conditions specified in the problem. This makes sense from the standpoint of conservation of charge. To see this, first note that $i_2 = 7$ A is leaving the node, while $i_1 = 2$ A is entering the node. To conserve charge, 7 A must be entering the node, which tells us that $i_2 = 5$ A is entering the node.

Kirchhoff's Voltage Law

The next fundamental tool we meet in circuit analysis is Kirchhoff's voltage law, which we abbreviate as KVL. This law tells us that at any instant of time in a loop in a circuit, the algebraic sum of the voltage drops in the circuit is zero.

$$\sum v(t) = 0 \quad \text{(for any loop in a circuit)} \tag{2.2}$$

We can consider a loop by moving clockwise or counterclockwise around the loop. To avoid mistakes, it is best to pick one way and stick to it. Consider the circuit shown in Fig. 2-4.

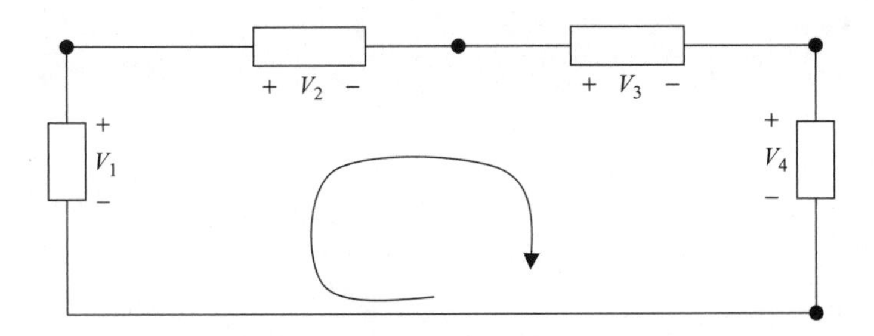

Fig. 2-4 A loop in a circuit. We add up voltage drops moving clockwise around the loop.

The circuit in Fig. 2-4 contains a single loop. Starting at V_1 and moving clockwise around the loop, we take the sign of each voltage to be positive or negative depending on whether we encounter a voltage drop or a voltage rise, respectively. Applying KVL to the circuit shown in Fig. 2-4 gives

$$-V_1 + V_2 + V_3 + V_4 = 0$$

To see how this works in practice, let's work two simple examples.

EXAMPLE 2-3
Consider the circuit shown in Fig. 2-5. Find the unknown voltage, V_x.

SOLUTION
Starting at the 5 V element, we will consider a clockwise loop around the circuit as indicated by the arrow drawn in the center. Voltage drops in the clockwise direction are positive, so

$$5 + 20 - 12 + V_x - 18 = 0$$

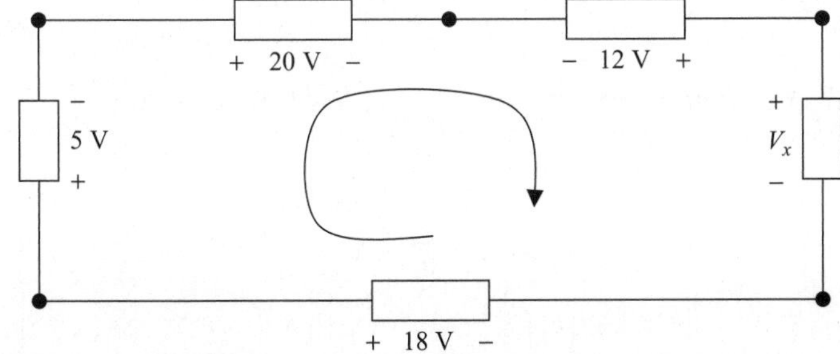

Fig. 2-5 Circuit solved in Example 2-3.

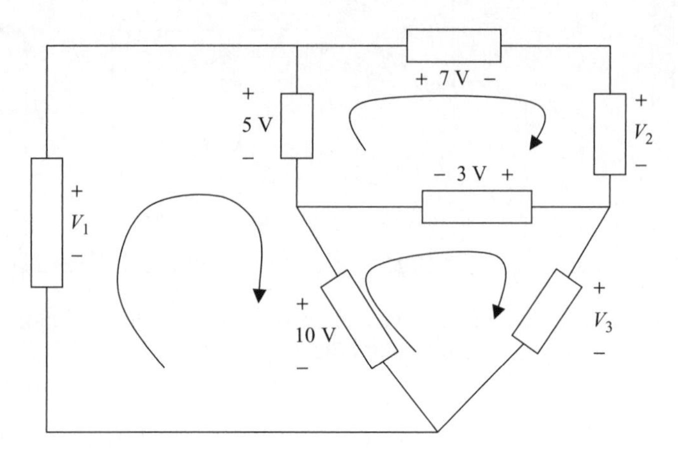

Fig. 2-6 The three-loop circuit of Example 2-4.

Solving for the unknown voltage

$$V_x = -5 - 20 + 12 + 18 = 5 \text{ V}$$

EXAMPLE 2-4
Consider the circuit shown in Fig. 2-6. Find the unknown voltages.

SOLUTION
To be consistent and therefore reduce our chances for error, we again consider clockwise loops, taking each voltage drop we encounter to be positive and each voltage rise we encounter to be negative. There are three loops in the circuit and we apply KVL to each loop individually. Starting on the left side of the circuit

$$-V_1 + 5 + 10 = 0$$
$$\Rightarrow V_1 = 5 + 10 = 15 \text{ V}$$

Next, moving to the loop on the top-right side of the circuit

$$-5 + 7 + V_2 + 3 = 0$$
$$\Rightarrow V_2 = 5 - 7 - 3 = -5 \text{ V}$$

Since $V_2 < 0$, the actual polarity is the opposite of what is shown in the figure, i.e., the actual $+/-$ signs are reversed for V_2. Continuing by considering the

loop in the lower-right part of the circuit

$$-10 - 3 + V_3 = 0$$
$$\Rightarrow V_3 = 10 + 3 = 13 \text{ V}$$

Now we've found all the voltages in the circuit, but we must note one other fact. KVL is a general law that applies to the voltages around *any* loop. KVL can therefore be applied to the outside loop in Fig. 2-6. Hence

$$-V_1 + 7 + V_2 + V_3 = 0$$

Is this true given the results we have found? Let's insert the numbers

$$-V_1 + 7 + V_2 + V_3 = -15 + 7 - 5 + 13 = -15 + 2 + 13 = 0$$

So KVL is indeed satisfied for the outer loop.

The Resistor

In this chapter we study our first circuit element in detail, the *resistor.* As we will see, the resistor is actually a very simple device, so our analysis won't change too much at this point. We will just have to do a bit of extra algebra.

The operation of a resistor is based on the following fact from physics. As we know, a current is a flow of charges—in other words, the charges in a material are moving in a given direction at some speed. As the charges move through the material, they are going to collide with atoms that are fixed in place in the form of a crystalline lattice. As the charges move, they follow a process whereby they gain speed, move some distance, then collide with an atom, and have to start all over again. To get them going we need to apply some kind of external force.

The external force is applied by impressing an electric field on the material. For small velocities the *current density* **J,** which is coulombs per cubic meter in SI units, is related to the electric field via a linear relation of the form

$$\mathbf{J} = \sigma \mathbf{E} \tag{2.3}$$

In short, an applied voltage (and hence electric field) gives the charges the energy they need to maintain their motion and keep the current going. The constant of proportionality, which we have denoted by σ, is the *conductance* of the material. The larger the σ is, the larger the current density **J** is for a given electric field. Metals such as copper or aluminum, which are good conductors, have very large values of σ, while a material like glass or wood will have a small value of σ.

The inverse of conductivity is *resistivity,* which is denoted by the Greek symbol ρ. Resistivity is the inverse of conductivity

$$\rho = \frac{1}{\sigma}$$

So we could write (2.3) as

$$\mathbf{E} = \rho \mathbf{J} \tag{2.4}$$

The units of resistivity are ohm-meters. However, in circuit analysis where we concern ourselves with lumped elements we are more interested in *resistance* (a lumped element is one that has no spatial variation of v or i over the dimensions of the element). The dimensions of the element will not be important; only the global properties of the element are of concern to us. We measure *resistance* in *ohms*, which are denoted by the upper case Greek character omega

$$\Omega \quad \text{(ohms)} \tag{2.5}$$

Resistance is usually denoted by R, which is a constant of proportionality between voltage and current. This relationship comes straight from (2.4) and is called *Ohm's law* after its discoverer. In terms of voltage and current, it is written as follows

$$V = RI \tag{2.6}$$

It is possible for resistance to vary with time, but in many if not most cases it is a constant. If the current and voltage vary with time, then Ohm's law can be written as

$$v(t) = Ri(t) \tag{2.7}$$

One way to think about what resistance does is to rewrite Ohm's law so that we have the current in terms of the voltage. That is (2.6) can be written as

$$I = \frac{V}{R}$$

Fig. 2-7 A schematic representation of a resistor.

With the equation in this form, we see that for a given applied voltage, if the resistance of the material is larger, the resulting current will be smaller.

The inverse of resistance is the *conductance*, G

$$G = \frac{1}{R} \text{ (siemens)} \tag{2.8}$$

As indicated, the SI unit of conductance is the siemens, a name that comes from a mysterious German scientist who studied electric properties of materials some time ago. While we will stick to SI units in this book, be aware that conductivity is also measured in *mhos*, which are denoted by an upside-down omega symbol.

We indicate a resistor in a circuit by drawing a jagged line, as shown in Fig. 2-7.

If a device resists the flow of current, the energy has to go somewhere. This is usually reflected in the emission of heat or light from the device. Resistance is found in many practical electric components and appliances. Perhaps the most familiar example of a resistor is the filament in a light bulb, where the resistance gives rise to light. Another example is a toaster, where resistive elements give off heat and some light that is useful to toast bread. In other cases, resistance might not be as useful; for example, an electric chord might have a bit of resistance that results in heat.

Power in a Resistor

The power absorbed or delivered by a resistor can be calculated from the expression $P = VI$ together with Ohm's law (2.6) $V = RI$. If we know the resistance and the voltage, then

$$P = VI = \frac{V^2}{R} = GV^2 \tag{2.9}$$

On the other hand, if we know the current through the resistor then we can write

$$P = VI = RI^2 \tag{2.10}$$

Remember, a resistor is an element that gives off energy, usually in the form of heat and sometimes in the form of light. Hence, a resistor *always absorbs* power.

Circuit Analysis with Resistors

When doing circuit analysis with a network that contains resistors, we apply KCL and KVL using Ohm's law to relate the voltage to the current as necessary. The best way to proceed is to look at some examples.

EXAMPLE 2-5
Find the three unknown currents shown in Fig. 2-8.

SOLUTION
We will denote the voltage across each resistor R by V_R. First we apply KVL to each of the two panes or loops in the circuit. Going in a clockwise direction, the loop on the left-hand side of Fig. 2-8 gives

$$-7 + V_5 = 0$$

where V_5 is the voltage across the 5 Ω resistor in the center. We conclude from KVL that $V_5 = 7$ V. Using Ohm's law (2.6) we can find the current through the

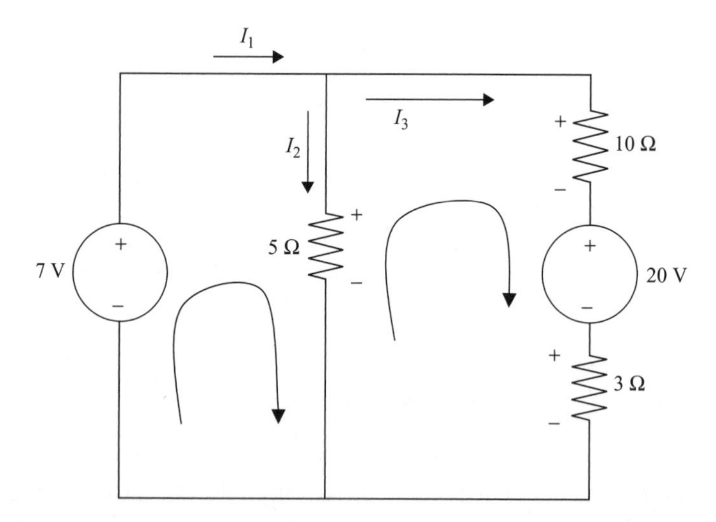

Fig. 2-8 Circuit analyzed in Example 2-5. Note the voltage polarities that have been specified for each resistor.

resistor, which is just I_2

$$I_2 = \frac{V_5}{5} = \frac{7}{5} = 1.4 \text{ A} \tag{2.11}$$

Next we apply KVL to the right-hand loop in Fig. 2-8. Again, we take the loop in a clockwise direction. This gives

$$-V_5 + V_{10} + 20 + V_3 = 0 \tag{2.12}$$

Although we know $V_5 = 7$ V, this equation leaves us with two unknowns. With one equation and two unknowns we need more information to solve the problem. Some extra information comes in the form of Ohm's law. The same current I_3 flows through the 10 and 3 Ω resistors. Hence $V_{10} = 10I_3$ and $V_3 = 3I_3$ and we can write (2.12) as

$$-7 + 10I_3 + 20 + 3I_3 = 0$$
$$\Rightarrow I_3 = -1 \text{ A}$$

Knowing two of the currents, we can solve for the other current by considering KCL at the top-center node. We take + for currents entering the node and – for currents leaving the node. This gives

$$I_1 - I_2 - I_3 = 0$$

Therefore we have

$$I_1 = I_2 + I_3 = 1.4 - 1 = 0.4 \text{ A}$$

Now we apply Ohm's law again to get the voltages across each of the resistors

$$V_{10} = 10I_3 = 10(-1) = -10 \text{ V}$$
$$V_3 = 3I_3 = 3(-1) = -3 \text{ V}$$

Notice the minus signs. These tell us that the actual voltages have the polarity opposite to that indicated in Fig. 2-8.

EXAMPLE 2-6

Let's consider a simple abstract model of a toaster. Our model will consist of the wall outlet, an electric chord, a switch, and a heating element. The wall outlet

is modeled as a voltage source given by

$$v_s(t) = 170 \sin 377t$$

The chord, which for our purposes transmits current to the toaster and dissipates heat, will be modeled as a resistor, which we denote as R_c. In this example

$$R_c = 20\ \Omega$$

We model the toaster by a resistor $R_t = 10\ \Omega$. Don't worry about the values of resistance given. These are just for instructional purposes so that you can get a feel of doing circuit analysis; they don't necessarily reflect realistic values. At time $t = 0$, the switch is closed and the toaster starts operation. Draw the circuit model and find the current that flows through the toaster together with the voltages across each resistor.

SOLUTION

We draw the elements as a series circuit, the wall outlet connected to the chord that is connected to the switch that is connected to the resistor representing the toaster. This is shown in Fig. 2-9.

To solve the problem, imagine the switch closing to make a complete circuit. There is only one loop to worry about in this circuit, so applying KVL in a

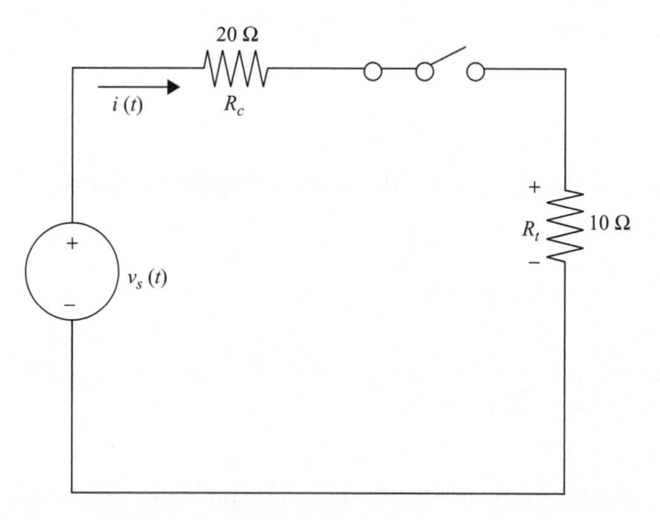

Fig. 2-9 A simple circuit model of a toaster.

clockwise loop that starts at the voltage source we obtain

$$-170 \sin 377t + v_c(t) + v_t(t)$$

where $v_c(t)$ is the voltage across the resistor representing the chord and $v_t(t)$ is the voltage across the resistor representing the toaster. With a single loop, it should be clear that the same current will flow between all components. We can use Ohm's law to write the voltages across the chord and toaster in terms of this current

$$v_c(t) = R_c i(t), \quad v_t(t) = R_t i(t)$$

Hence KVL becomes

$$-170 \sin 377t + 20i(t) + 10i(t) = 0$$

$$\Rightarrow i(t) = \frac{170}{20 + 10} \sin 377t = 5.7 \sin 377t$$

The voltage across the chord resistor is (using Ohm's law)

$$v_c(t) = R_c i(t) = (20)(5.7) \sin 377t = 114 \sin 377t$$

And the voltage across the resistor representing the toaster is

$$v_t(t) = R_t i(t) = (10)(5.7) \sin 377t = 57 \sin 377t$$

Root Mean Square (RMS) Values

Many electric appliances that you are familiar with, including the standard household outlet, do not list the actual time varying voltages and currents. Instead, they list the *effective value* of the current or voltage which is commonly called the *root mean square* or *RMS* value. This quantity is defined as follows. What constant or DC source would produce the same *average* power as the actual time varying source? This is the RMS voltage or current and it can be calculated by using the following three steps:

- Square the time varying voltage or current
- Find the average over one period
- Take the positive square root

When calculating RMS values, the following information can be helpful. The *total energy* w over some time interval $a \le t \le b$ is found by integrating the power $p(t)$

$$w = \int_a^b p(t)\,dt \tag{2.13}$$

The units of (2.13) are joules, since power is measured in joules/second. The *average power* over the interval $a \le t \le b$ is found by dividing the total energy by the interval

$$\text{Average Power} = \frac{w}{b-a} \tag{2.14}$$

To summarize, the effective or RMS voltage is the constant voltage that would produce the *same average power* as the actual voltage. Using the power given by Ohm's law (2.9) we have

$$G V_{\text{RMS}}^2 = \frac{\int v(t)\,dt}{\Delta t} G \tag{2.15}$$

where G is the conductivity.

When someone says that the household outlet is 120 V, they are quoting the RMS voltage—the actual voltage is $v(t) = 170 \sin 377t$. Likewise the RMS current on an electric appliance is the constant current that would produce the same average power as the actual current.

EXAMPLE 2-7

Show that the RMS voltage for a household outlet is 120 V.

SOLUTION

We start with $v(t) = 170 \sin 377t$. Step one is to square this voltage

$$v^2(t) = (170)^2 \sin^2 377t$$

To integrate this quantity, we use a trig identity to rewrite it

$$\sin^2 \omega t = \frac{1 - \cos 2\omega t}{2}$$

To find the energy, we integrate over one period. The frequency is found for a sin wave using the relationship $\omega = 2\pi\nu$, where ν is the frequency in hertz (Hz) or cycles per second. In this case

$$\nu = \frac{377}{2\pi} = 60 \text{ Hz}$$

The period T is the inverse of this quantity

$$T = \frac{1}{\nu} = \frac{1}{60} \text{ s}$$

Therefore we integrate

$$\int_0^T v^2(t)\,dt = (170)^2 \int_0^{1/60} \left(\frac{1 - \cos\left[2\,(377t)\right]}{2} \right) dt$$

We don't have to worry about the cosine term, since

$$\int_0^T \cos(2\omega t)\,dt = \frac{1}{2\omega} \sin 2\omega t \Big|_0^T = \frac{1}{2\omega} \sin 2\omega T - \frac{1}{2\omega} \sin(0) = \frac{1}{2\omega} \sin 2\omega T$$

However using the definition of period in terms of frequency this term vanishes

$$\frac{1}{2\omega} \sin 2\omega T = \frac{1}{2\omega} \sin\left(2\omega \frac{2\pi}{\omega}\right) = \frac{1}{2\omega} \sin 4\pi = 0$$

So the energy over one cycle is

$$\int_0^T v^2(t)\,dt = (170)^2 \int_0^{1/60} \left(\frac{1 - \cos\left[2\,(377t)\right]}{2} \right) dt$$

$$= \frac{(170)^2}{2} \int_0^{1/60} dt = \frac{(170)^2}{2} \left(\frac{1}{60} \right)$$

To calculate the average power over a cycle, we set Δt in (2.15) equal to the time of one cycle, which is 1/60 of a second. Therefore the average power is

$$\frac{\frac{(170)^2}{2}\left(\frac{1}{60}\right)}{1/60}G = \frac{(170)^2}{2}G$$

We can cancel the conductivity in (2.15) to solve for the RMS voltage

$$V_{\text{rms}} = \sqrt{\frac{(170)^2}{2}} = \frac{170}{\sqrt{2}}\text{ V} = 120\text{ V}$$

This example demonstrates a useful trick. It's not necessary to go through the integral when the source is a sinusoidal voltage or current. The effective or RMS voltage or current for a sinusoidal source is found by dividing the amplitude of the source by the square root of two. That is, if $v(t) = V_m \sin \omega t$, then

$$V_{\text{rms}} = \frac{V_m}{\sqrt{2}} \tag{2.16}$$

The average power loss for a resistor R given this source is

$$P_{\text{av}} = \frac{V_{\text{rms}}^2}{R} = \frac{V_m^2}{2R} \tag{2.17}$$

The RMS current is related to the amplitude of a sinusoidal current via

$$I_{\text{rms}} = \frac{I_m}{\sqrt{2}} \tag{2.18}$$

And the average power is

$$P_{\text{av}} = RI_{\text{rms}}^2 = \frac{RI_m^2}{2} \tag{2.19}$$

Remember that the effective voltage or current means the same thing as the RMS voltage or current.

EXAMPLE 2-8

Find the effective current and voltage for the toaster used in Example 2-6. What is the average power?

SOLUTION

The current through the resistor representing the toaster was found to be

$$i(t) = 5.7 \sin 377t$$

Therefore the amplitude is

$$I_m = 5.7 \text{ A}$$

The current is sinusoidal. Hence we can apply (2.18)

$$I_{\text{eff}} = \frac{I_m}{\sqrt{2}} = \frac{5.7}{\sqrt{2}} = 4 \text{ A}$$

The effective voltage can be calculated by using (2.16) with $v_t(t) = 57 \sin 377t$

$$V_{\text{eff}} = \frac{V_m}{\sqrt{2}} = \frac{57}{\sqrt{2}} = 40 \text{ V}$$

The average power can be calculated by using (2.19). In Example 2-6, we were told that $R_t = 10\ \Omega$. So we find the average power to be

$$P_{\text{av}} = RI_{\text{eff}}^2 = (10)(4)^2 = 160 \text{ W}$$

Voltage and Current Dividers

If a set of resistors is connected in a series, the voltage across any resistor can be calculated without having to know the current. This is because the same current flows through any set of resistors connected in a series. The *equivalent resistance* or total resistance of a set of resistors connected in a series is found by summing up the resistances of each of the individual components. That is,

$$R_{\text{eq}} = \sum R_i \quad \text{(for resistors in series)} \tag{2.20}$$

If the resistors in a series are connected to a voltage source $v_s(t)$, then the current is

$$i(t) = \frac{v_s(t)}{R_{\text{eq}}} \tag{2.21}$$

To find the voltage across the jth resistor R_j we apply the *voltage division* or *voltage divider* rule

$$v_j(t) = \frac{R_j}{R_{eq}} v_s(t) \tag{2.22}$$

If two resistors are connected in a series to a source, then the voltage across resistor 1 is

$$v_1(t) = \frac{R_1}{R_1 + R_2} v_s(t)$$

while the voltage across the second resistor is

$$v_2(t) = \frac{R_2}{R_1 + R_2} v_s(t)$$

EXAMPLE 2-9

Find the voltage across the second resistor in the circuit shown in Fig. 2-10.

SOLUTION

The circuit is shown in Fig. 2-10 with DC voltage source $V_s = 8$ V.

The equivalent resistance is

$$R_{eq} = 1 + 3 + 4 = 8\ \Omega$$

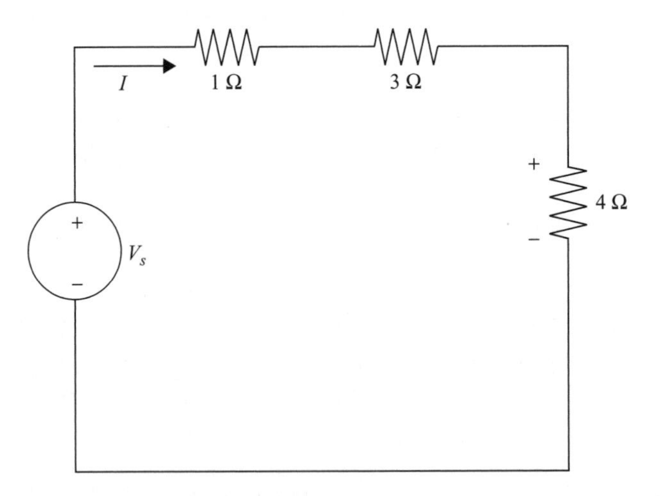

Fig. 2-10 The voltages in this circuit can be found by using voltage dividers.

The voltage across the second resistor in the series is

$$V_2 = \frac{R_2}{R_{\text{eq}}} V_s = \left(\frac{3\ \Omega}{8\ \Omega}\right) 8\ \text{V} = 3\ \text{V}$$

What is the power delivered by the voltage source? The current flowing through the circuit can be found using Ohm's law

$$I = \frac{3\ \text{V}}{3\ \Omega} = 1\ \text{A}$$

The power is

$$P = VI = (8\ \text{V})(1\ \text{A}) = 8\ \text{W}$$

Many times in a circuit diagram you will see a *ground* or *reference node.* All voltages in a circuit are taken to be positive with respect to this node, which is usually placed at the bottom of the circuit diagram. The symbol used for a ground node is shown in Fig. 2-11.

Now consider a set of resistors in parallel, as shown in Fig. 2-12. Using KVL, it is easy to convince yourself that the same voltage V is found across each resistor. What is the value of that voltage?

The sum of the currents leaving the first node in Fig. 2-12 must be equal to the current source

$$I_s = I_1 + I_2 + I_3 = \frac{1}{R_1} V + \frac{1}{R_2} V + \frac{1}{R_3} V = \left(\frac{1}{R_1} + \frac{1}{R_2} + \frac{1}{R_3}\right) V$$

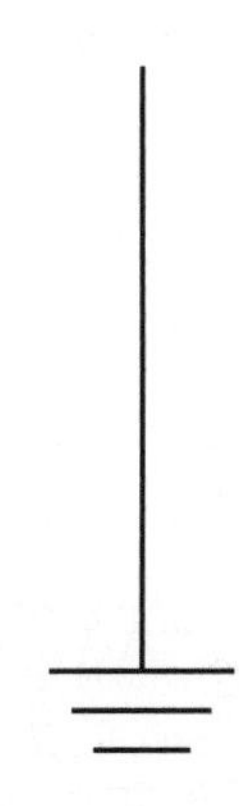

Fig. 2-11 A ground or reference symbol.

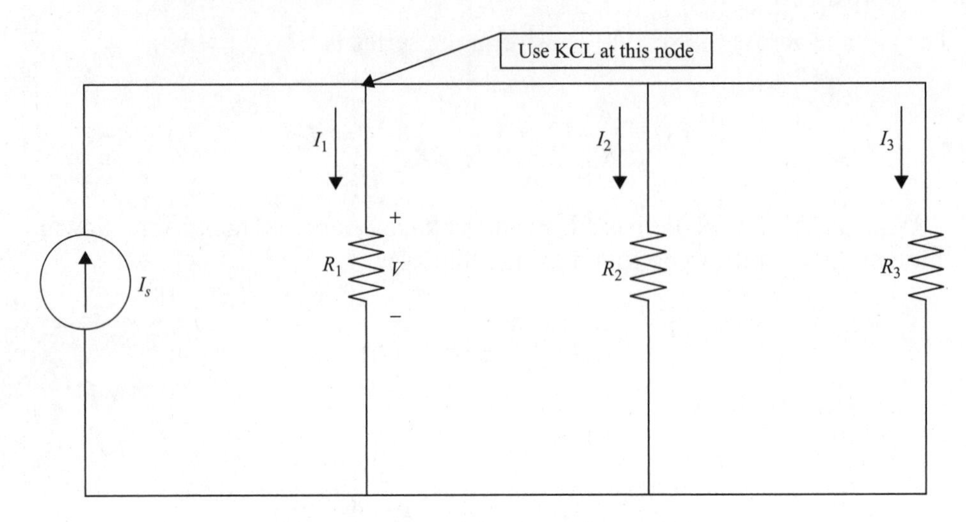

Fig. 2-12 A set of resistors in parallel.

We applied Ohm's law at the last step and used the fact that the same voltage is across each resistor, which you can verify using KVL. The inverse of the resistance is the conductance and so

$$I_s = (G_1 + G_2 + G_3)V$$

This is an example of total or *equivalent conductance*, which is just the sum of conductances connected in parallel

$$G_{\text{eq}} = \sum G_i \tag{2.23}$$

EXAMPLE 2-10

What is the total resistance of the circuit shown in Fig. 2-13, and what is the voltage across each resistor?

SOLUTION

The total resistance is

$$R_T = \frac{1}{1/R_1 + 1/R_2} = \frac{1}{1/2 + 1/4} = \frac{4}{3}\ \Omega$$

A rule of thumb is that the total resistance of two resistors in parallel is

$$R_T = \frac{R_1 R_2}{R_1 + R_2}$$

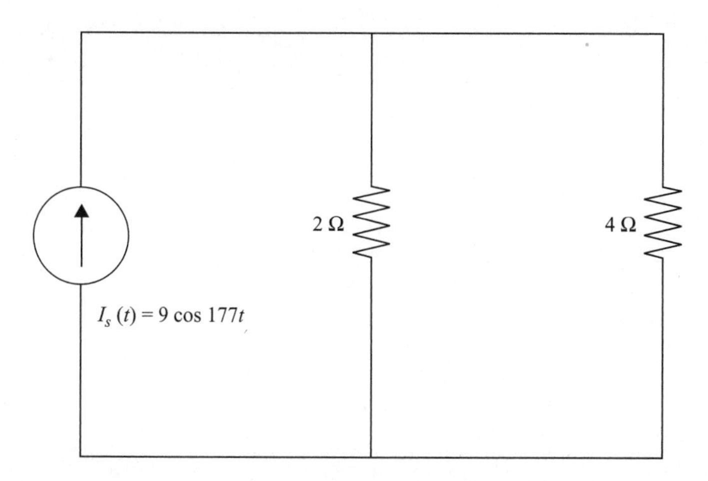

Fig. 2-13 The circuit used in Example 2-10.

The voltage across each resistor can be found using the fact that $i_s(t) = G_T v(t)$, from which we conclude that $v(t) = R_T i_s(t)$, which gives

$$v(t) = \left(\frac{4}{3}\right) 9 \cos 177t = 12 \cos 177t \text{ V}$$

This example brings us to the concept of a *current divider*. To find the current flowing through an individual resistor when several resistors are connected in parallel, we use the current divider rule, which says

$$I_j = \frac{G_j}{G_{\text{eq}}} I_s \tag{2.24}$$

where I_j is the current through the jth resistor, G_j is the resistor's conductivity, G_{eq} is the equivalent conductivity of the circuit, and I_s is the source current. For the special case of two resistors

$$I_1 = \frac{R_2}{R_1 + R_2}, \quad I_2 = \frac{R_1}{R_1 + R_2}$$

EXAMPLE 2-11

For the circuit shown in Fig. 2-14, find the current flowing through each resistor and the voltage across each resistor.

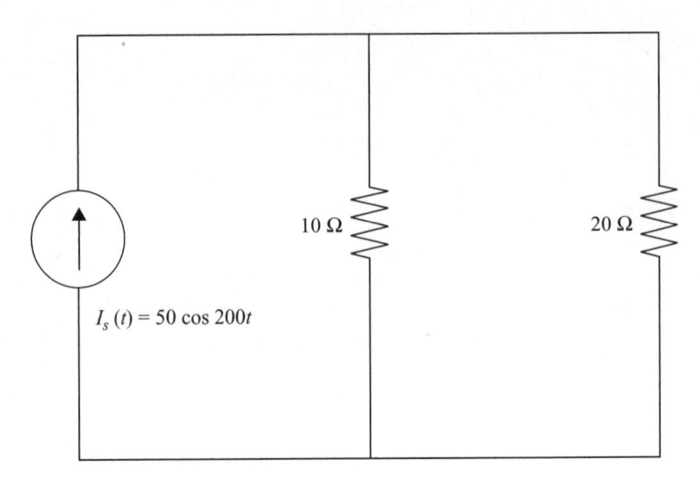

Fig. 2-14 Circuit used in Example 2-11, illustrating the use of a current divider.

SOLUTION

Using the formula for the special case of (2.24) the current in the first resistor is

$$i_1(t) = \frac{R_2}{R_1 + R_2} i_s(t) = \left(\frac{20}{10 + 20}\right) 50 \cos 200t = 33.3 \cos 200t \text{ A}$$

The current flowing in the second resistor is

$$i_2(t) = \frac{R_1}{R_1 + R_2} i_s(t) = \left(\frac{10}{10 + 20}\right) 50 \cos 200t = 16.7 \cos 200t \text{ A}$$

Notice that $i_1 + i_2 = i_s$, as required by the conservation of charge. The voltage across both resistors is the same (check it by using KVL) and can be found by applying Ohm's law to either resistor. Choosing the 10 Ω resistor

$$v(t) = (10\ \Omega)(33.3 \cos 200t \text{ A}) = 333 \cos 200t \text{ V}$$

More Examples

We conclude the chapter with more examples that illustrate the use of KCL, KVL, and basic resistive circuits.

EXAMPLE 2-12

A high-voltage transmission line with a resistance of 0.065 Ω/mi distributes power to a load 220 miles away. Modeling the load as a resistor, find the load resistance R_L such that the power at the load is 500 MW. What percentage of power generated by the source is "wasted" as heat dissipated by the transmission line? The power source is $V_s = 300$ kV.

SOLUTION

The high-voltage transmission line can be modeled as a resistor R_T. The total resistance of the line is

$$R_T = (0.065\ \Omega/\text{mile})(220\ \text{miles}) = 14\ \Omega$$

The model for the entire system is actually very simple. The system can be modeled by a circuit consisting of the source, the transmission line, and load all connected in a series. This is shown in Fig. 2-15.

Voltage dividers can be used to give the voltage across each resistor

$$V_T = \frac{R_T}{R_T + R_L} V_s, \quad V_L = \frac{R_L}{R_T + R_L} V_s$$

The power across the load is

$$P_L = \frac{V_L^2}{R_L} \tag{2.25}$$

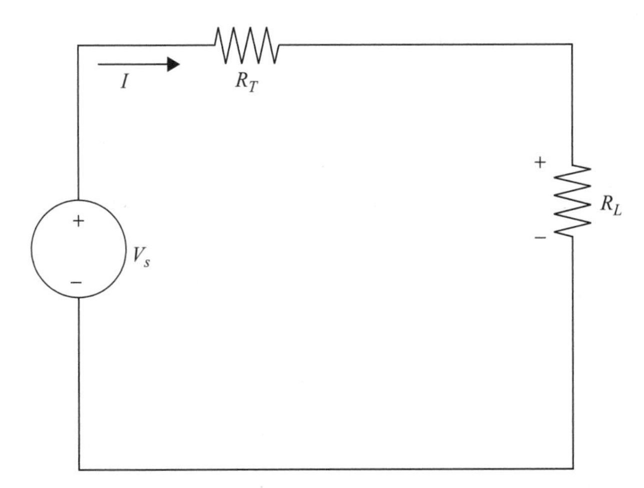

Fig. 2-15 A model of high-power transmission line and load.

Using the voltage divider we can reduce this to a single unknown in terms of the load resistance

$$P_L = \frac{V_L^2}{R_L} = \frac{1}{R_L}\frac{R_L^2}{(R_T + R_L)^2}V_s^2 = \frac{R_L}{(R_T + R_L)^2}V_s^2$$

Expanding $(R_T + R_L)^2$ and rearranging terms we get a quadratic equation in terms of the load resistance

$$R_L^2 + \left(2R_T - \frac{V_s^2}{P_L}\right) + R_T^2 = 0 \tag{2.26}$$

No doubt the reader recalls we can solve this equation using the quadratic formula

$$\frac{-b \pm \sqrt{b^2 - 4ac}}{2a}$$

In (2.26), $a = 1,\ b = 2R_T - \frac{V_s^2}{P_L}$, and $c = R_T^2$. Putting the numbers in

$$b = 2R_T - \frac{V_s^2}{P_L} = 2(14) - \frac{(300\ \text{kV})}{500\ \text{MW}} = -15.2\ \Omega \tag{2.27}$$

$$c = R_T^2 = (14)^2 = 196 \tag{2.28}$$

Putting the numbers into the quadratic formula we find that the load resistance is

$$R_L = 7.6 \pm \frac{\sqrt{(-15.2)^2 - 4(196)}}{2} = 7.6 \pm 75\ \Omega$$

Resistance is always positive, so we take the + sign and find

$$R_L = 82\ \Omega$$

Using voltage dividers, the voltage across the transmission line is

$$V_T = \frac{R_T}{R_T + R_L}V_s = \frac{14}{14 + 82}300 = 44\ \text{kV}$$

The voltage across the load is

$$V_L = \frac{R_L}{R_T + R_L} V_s = \frac{82}{14 + 82} 300 = 256 \text{ kV}$$

Notice that conservation of energy is satisfied and that the sum of these terms is equal to the voltage supplied by the source. The current through the circuit is

$$I = \frac{V_L}{R_L} = \frac{256 \text{ kV}}{82 \ \Omega} = 3.12 \text{ kA}$$

Pay attention to the units! The power of the source can now be calculated using $P = VI$

$$P_s = V_s I = (300 \text{ kV})(3.12 \text{ kA}) = 937 \text{ MW}$$

The power at the load is

$$P_L = \frac{V_L^2}{R_L} = \frac{(256 \text{ kV})^2}{82 \ \Omega} = 799 \text{ MW}$$

The power dissipated at the transmission line is

$$P_T = \frac{V_T^2}{R_T} = \frac{(44 \text{ kV})^2}{14 \ \Omega} = 138 \text{ MW}$$

The amount of power wasted as a percentage as heat emitted by the transmission line is

$$\% = \frac{P_T}{P_s} \times 100 = \frac{138}{936} \times 100 = 14.7\%$$

EXAMPLE 2-13
Find the power dissipated in the 3 Ω resistor shown in Fig. 2-16.

SOLUTION
Applying KVL to the left pane using a clockwise loop

$$-24 + 6I = 0$$

$$\Rightarrow I = \frac{24}{6} = 4 \text{ A}$$

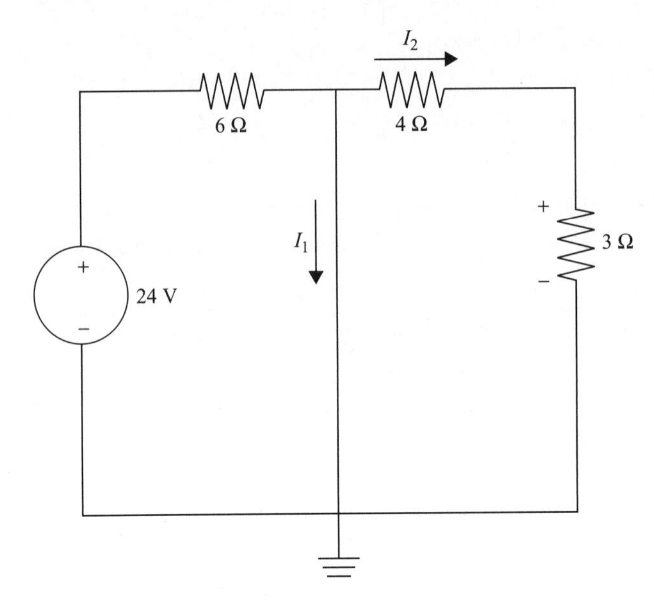

Fig. 2-16 What is the effect of a short circuit in parallel with two resistors?

KCL at the top-middle node (directly above ground) taking + for currents entering and − for currents leaving gives

$$I - I_1 - I_2 = 0 \Rightarrow I_1 + I_2 = I$$

KVL around the outside loop gives

$$-24 + (6)(4) + 4I_2 + 3I_2 = 0$$
$$\Rightarrow I_2 = 0$$

No current flows through the 3 Ω resistor; hence the power dissipated is zero. The short circuit draws all of the current. This isn't surprising since it has zero resistance.

EXAMPLE 2-14

Recalling Cramer's rule from linear algebra, find the unknown currents for the circuit in Fig. 2-17. Suppose that $R_1 = 10\ \Omega,\ R_2 = 4\ \Omega,\ R_3 = 1\ \Omega$ and $V_1 = 10$ V, $V_2 = 3$ V, and $V_3 = 6$ V.

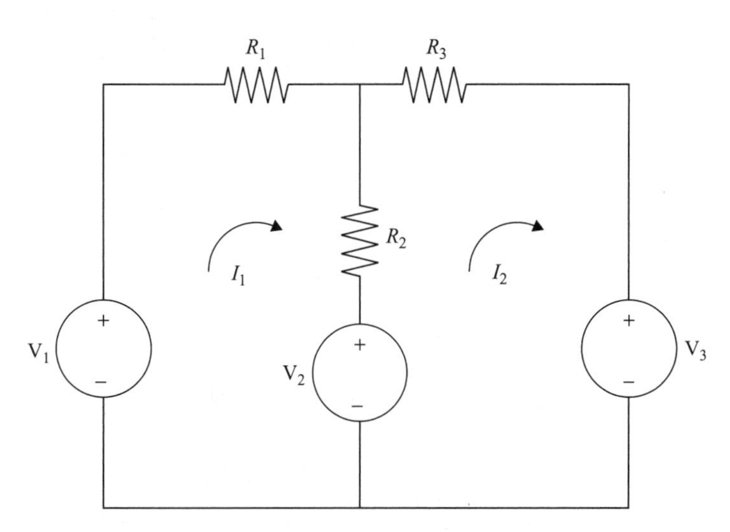

Fig. 2-17 In Example 2-14 we solve a mesh problem by using Cramer's rule.

SOLUTION

We apply KVL to each loop, but we must be careful with the middle resistor. Notice that the currents through the middle resistor are moving in opposite directions, as shown here

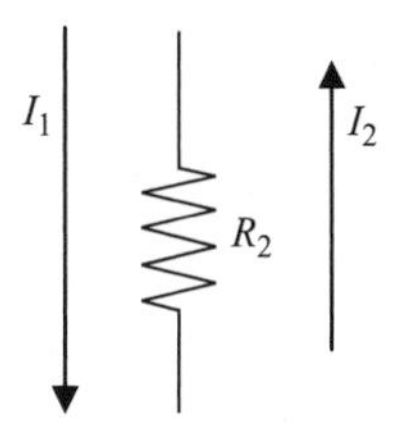

To apply KVL, we use the loop current as the positive sense, so the current through R_2 in the left pane is $I_1 - I_2$, but the current in the right pane is taken to be $I_2 - I_1$, because the current I_2 sets the positive direction in the right-hand pane. With this in mind, applying KVL in the left pane gives

$$-V_1 + R_1 I_1 + R_2(I_1 - I_2) + V_2 = 0$$

Rearranging a bit gives

$$(R_1 + R_2)I_1 - R_2 I_2 = V_1 - V_2 \tag{2.29}$$

Applying KVL to the right side we get

$$-V_2 + R_2(I_2 - I_1) + R_3 I_2 + V_3 = 0$$

Cleaning up allows us to write this as

$$-R_2 I_1 + (R_2 + R_3) I_2 = V_2 - V_3 \tag{2.30}$$

Now putting in $R_1 = 10\ \Omega,\ R_2 = 4\ \Omega,\ R_3 = 1\ \Omega$ and $V_1 = 10$ V, $V_2 =$ 3 V, and $V_3 = 6$ V into (2.29) and (2.30), we obtain the following set of equations

$$14I_1 - 4I_2 = 7$$
$$-4I_1 + 5I_2 = -3$$

We can arrange the coefficients in a matrix and use Cramer's rule to solve for the currents. First we put the coefficients of the terms involving the currents on the left sides of each equation into a determinant

$$D = \begin{vmatrix} 14 & -4 \\ -4 & 5 \end{vmatrix} = (14)(5) - (-4)(-4) = 70 + 16 = 86$$

Next, we substitute the right-hand side, which is the column

$$\begin{pmatrix} 7 \\ -3 \end{pmatrix}$$

into the appropriate column to get the answer for I_1 and I_2. To get the answer for I_1, we replace the first column in D with the right side to obtain

$$D_1 = \begin{vmatrix} 7 & -4 \\ -3 & 5 \end{vmatrix} = (7)(5) - (-3)(-4) = 35 - 12 = 23$$

Then the current is

$$I_1 = \frac{D_1}{D} = \frac{23}{86} = 0.27\ \text{A}$$

Following the same procedure for the second current, we replace the second column of D to define

$$D_2 = \begin{vmatrix} 14 & 7 \\ -4 & -3 \end{vmatrix} = (14)(-3) - (-4)(7) = -42 + 28 = -14$$

And so the second current is

$$I_2 = \frac{D_2}{D} = \frac{-14}{86} = -0.16 \text{ A}$$

Summary

In this chapter we learned how to use Kirchhoff's current law (KCL) and Kirchhoff's voltage law (KVL) to solve for unknown quantities in circuits. These laws are basic principles that are based on the conservation of charge and energy, respectively, and they apply no matter what elements are used to construct the circuit. We then learned that a resistor is an element with the linear relation $V = RI$ between voltage and current that dissipates power as heat and/or light. We concluded the chapter by considering power in resistors and how to apply KVL and KCL to solve basic resistive circuits.

To review, Kirchhoff's current law (KCL) tells us that the sum of the currents at a node is zero:

$$\sum i(t) = 0$$

Kirchhoff's voltage law (KVL) tells us that the sum of the voltages around any loop in the circuit is zero:

$$\sum v(t) = 0 \quad \text{(for any loop in a circuit)}$$

Using Ohm's law, the power in a circuit element can be determined in terms of voltage

$$P = VI = \frac{V^2}{R} = GV^2$$

Or current:

$$P = VI = RI^2$$

For a sinusoidal voltage source $v(t) = V_m \sin \omega t$, the root mean square, RMS or effective voltage is given by:

$$V_{\text{rms}} = \frac{V_m}{\sqrt{2}}$$

While the average power delivered to a resistor R is:

$$P_{\text{av}} = \frac{V_{\text{rms}}^2}{R} = \frac{V_m^2}{2R}$$

The RMS current is:

$$I_{\text{rms}} = \frac{I_m}{\sqrt{2}}$$

And the power in terms of RMS current is:

$$P_{\text{av}} = RI_{\text{rms}}^2 = \frac{RI_m^2}{2}$$

Next we considered resistors in a series, which can be added to obtain an equivalent resistance:

$$R_{\text{eq}} = \sum R_i \quad \text{(for resistors in series)}$$

Finally, for resistors in parallel, we consider the conductance $G = 1/R$. Then we can obtain an equivalent conductance for a circuit by summing these up, for example:

$$I_s = (G_1 + G_2 + G_3)V$$

Where:

$$R_{\text{eq}} = \frac{1}{G_1 + G_2 + G_3}$$

Quiz

1. Consider the node shown in Fig. 2-18. Find the unknown current if $i_2 = -7$ A and $i_3 = 4$ A.

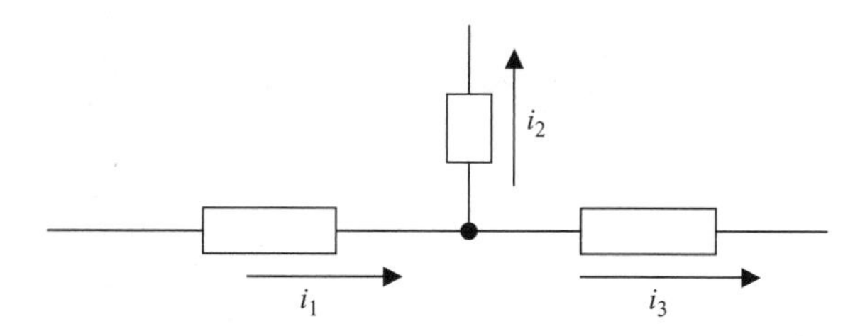

Fig. 2-18 Circuit for Problem 1.

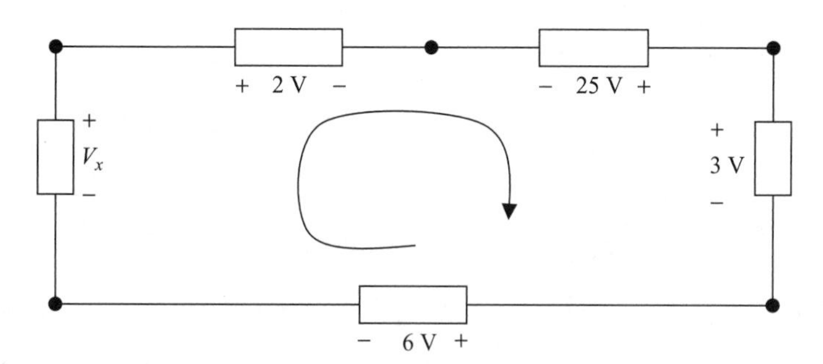

Fig. 2-19 Circuit for Problem 2.

2. Consider the circuit in Fig. 2-19, and find the unknown voltage.
3. Consider the circuit in Fig. 2-20. Find the unknown voltages.
4. It is known that the voltage across a resistor is 20 V, while 4 A of current flows through the resistor. What is the resistance?

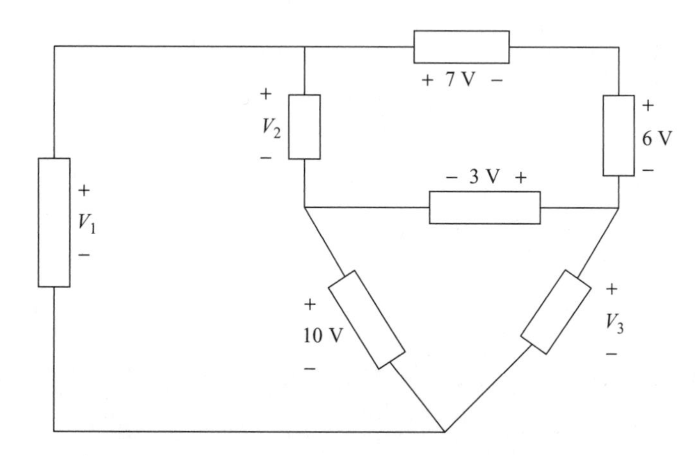

Fig. 2-20 Circuit for Problem 3.

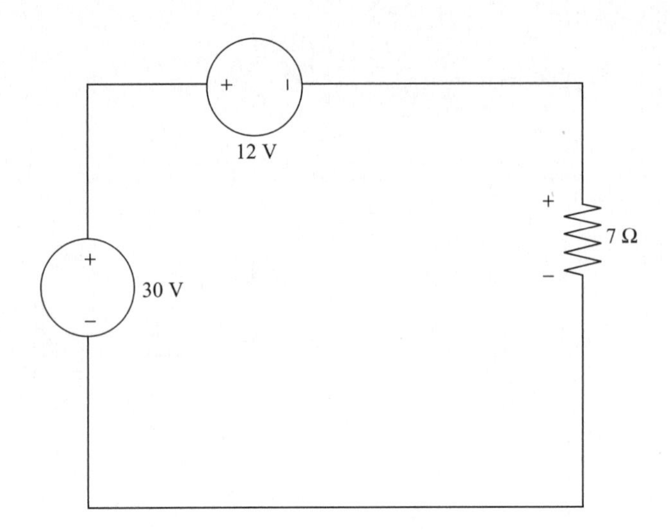

Fig. 2-21 Circuit for Problem 6.

5. In a circuit, 20 A of current flows through a 5 Ω resistor. What is the voltage? What is the conductance of the resistor?
6. In the circuit shown in Fig. 2-21, find the power dissipated or absorbed in the 7 Ω resistor.
7. Find the power dissipated in the 6 Ω resistor shown in Fig. 2-22.

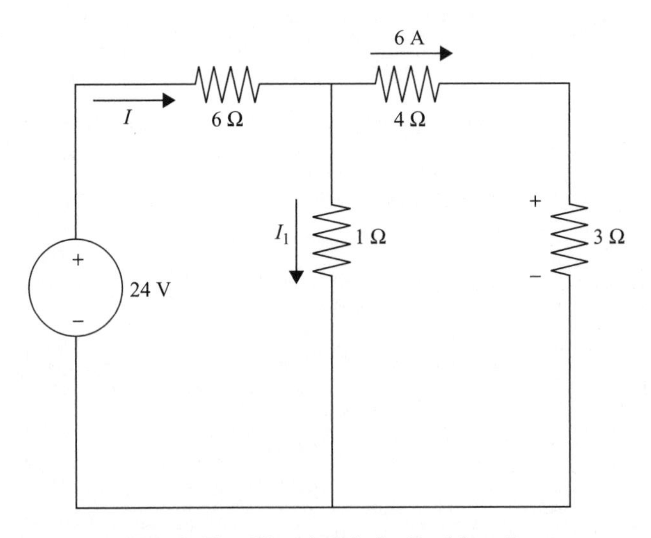

Fig. 2-22 The circuit for Problem 7.

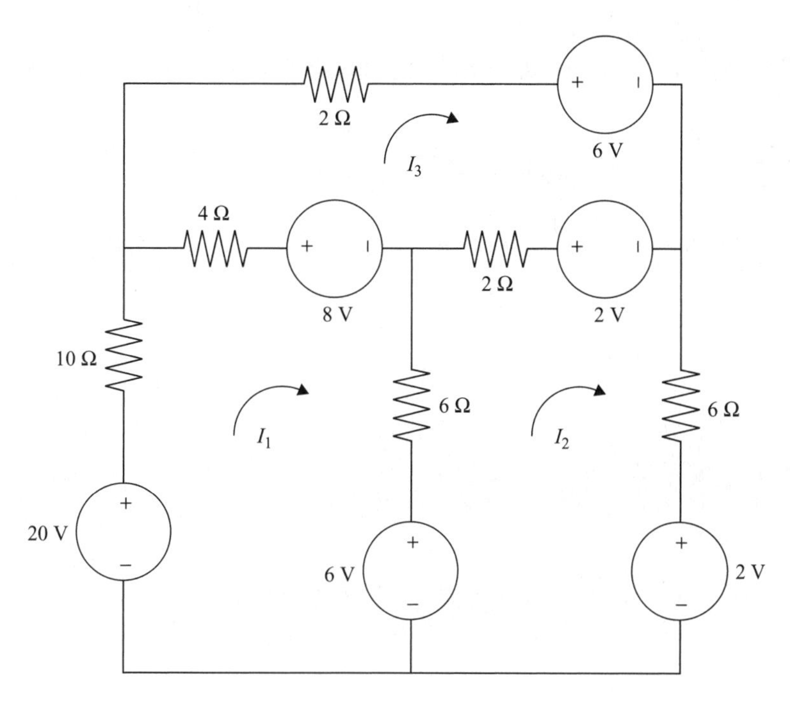

Fig. 2-23 Circuit for Problem 10.

8. A voltage source $v(t) = 30 \sin 10t$ is connected in series with a 5 Ω resistor, a 10 Ω resistor, and a voltage source $v(t) = 45 \sin 10t$. What is the current through the circuit?
9. Two resistors are connected in a series to a voltage source. The voltage across the first resistor is?
10. Using Cramer's rule, find the currents in the circuit shown in Fig. 2-23.

CHAPTER 3

Thevenin's and Norton's Theorems

Thevenin's and Norton's theorems are two techniques that allow us to simplify electric circuits. Consider a circuit that can be connected to the outside world (i.e., to other circuits) via two terminals that we label A and B that consist of voltage sources and resistors. Thevenin's theorem allows us to convert a complicated network into a simple circuit consisting of a single voltage source and a single resistor connected in series. The circuit is equivalent in the sense that it looks the same from the outside, that is, it behaves the same electrically as seen by an outside observer connected to terminals A and B. We will begin the chapter by introducing the traditional method used to apply Thevenin's theorem. After a few examples we will demonstrate a more recently derived way to apply Thevenin's theorem that relies on the introduction of a current source.

Norton's theorem is similar in that it allows us to replace a complicated electric circuit consisting of voltage sources and resistors with an equivalent

circuit consisting of a single current source and a resistor in parallel. Again, the circuit is connected to the outside world via two terminals A and B, and as far as the outside world is concerned the Norton equivalent circuit behaves in exactly the same way as the original circuit.

In summary, Thevenin's theorem and Norton's theorem are two techniques that we can call upon to simplify electric networks, as long as the network in question involves only resistors and voltage sources.

Thevenin's Theorem

Imagine that a certain electric circuit can be divided into two separate networks connected at terminals we label A and B as shown in Fig. 3-1.

On the left side of Fig. 3-1, we have a circuit consisting of resistors and voltage sources. It may be some complicated network of elements connected in various parallel and series combinations. On the right side of Fig. 3-1 is an outside network that is completely arbitrary. It is connected to the complicated network via two terminals labeled A and B, and our only concern is how the outside network "sees" the complicated network electrically. In other words, what are the voltage, current, and resistance at terminals A and B?

Thevenin's theorem tells us that, as far as the outside network is concerned, the circuit on the left can be replaced with a single resistor and a single current source. The resistor is denoted R_{TH} for Thevenin equivalent resistance and the voltage source is denoted by V_{TH} for Thevenin equivalent voltage. It does not matter if the voltage is constant, time varying, or sinusoidal, we can apply the theorem in each case.

The task at hand when applying Thevenin's theorem is to determine the value of the Thevenin equivalent voltage and the value of the Thevenin equivalent

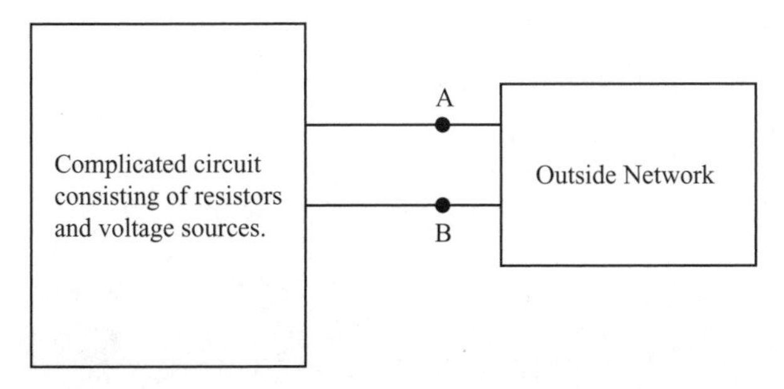

Fig. 3-1 A complicated circuit connected to an outside network at terminals A and B.

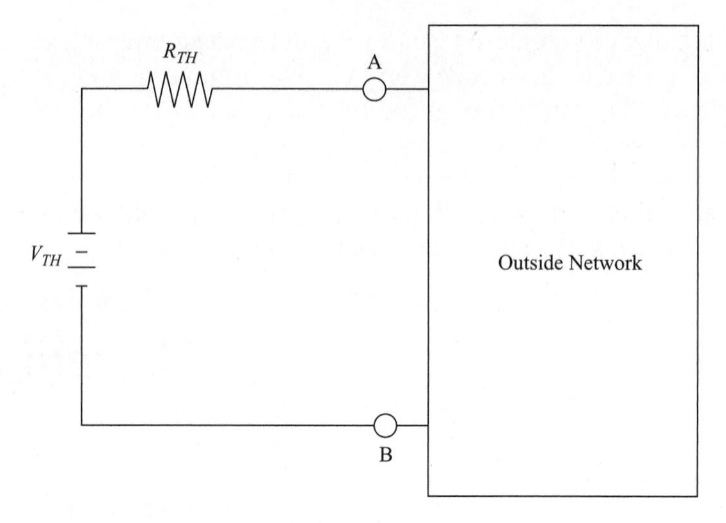

Fig. 3-2 Thevenin's theorem allows us to replace a network consisting of voltage sources and resistors in arbitrary connections by a single voltage source and resistor connected in series.

resistance. This can be done by applying the steps outlined in the following sections.

Step One: Disconnect the Outside Network

The first step in the application of Thevenin's theorem is to completely detach the outside network from terminals A and B. We then calculate the voltage across the open circuit at the two terminals A and B; this is the Thevenin equivalent voltage V_{TH}. It is illustrated in Fig. 3-3.

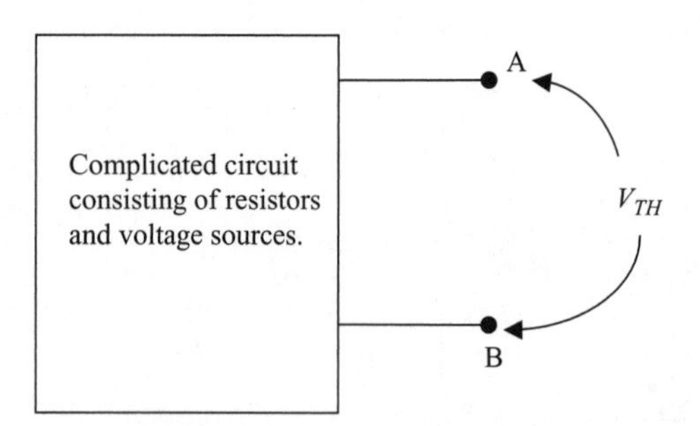

Fig. 3-3 The Thevenin equivalent voltage is calculated by finding the voltage across the open circuit at terminals A and B.

Step Two: Set Independent Sources to Zero

Next, we set all independent sources in the circuit to zero. Voltage sources and current sources are handled in the following way.

- For a voltage source v, set $v = 0$ and replace the voltage source by a short circuit.
- For a current source i set $i = 0$ and replace the current source by an open circuit.

If the original circuit contains any *dependent* sources, leave them unchanged in the circuit.

Step Three: Measure the Resistance at Terminals A and B

The final step in the application of Thevenin's theorem is to analyze the circuit with all independent sources set to zero and determine the resistance across terminals A and B. This is the Thevenin equivalent resistance R_{TH}. Finally, with the Thevenin equivalent voltage and resistance in hand, we draw the circuit with the voltage source and resistor in series as we did in Fig. 3-2.

At this point you may be confused, so we will illustrate the method by applying it to a few examples. First, however, we need to review the concept of resistors connected in series and in parallel.

Series and Parallel Circuits

Consider a set of n resistors all connected in a row, which is said formally by saying they are connected in *series* as shown in Fig. 3-4.

Since the resistors are connected in series, the same current I flows through each resistor. Applying Ohm's law, this means that the voltage through

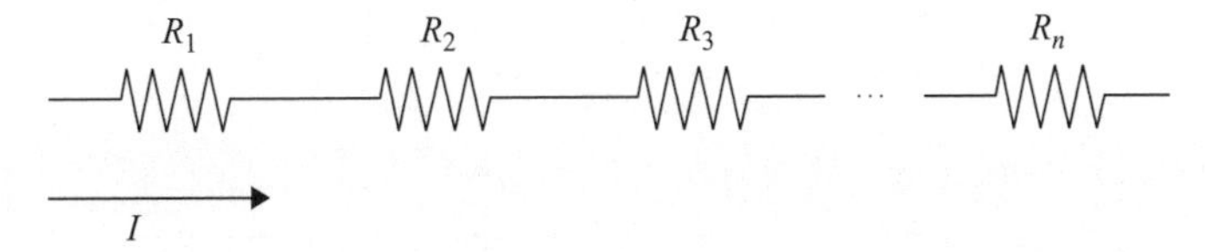

Fig. 3-4 A set of resistors connected in series.

each resistor is

$$V_1 = R_1 I, \quad V_2 = R_2 I, \quad V_3 = R_3 I, \ldots, \quad V_n = R_n I$$

The total voltage across the entire set is found by adding up the voltage across each resistor

$$V = V_1 + V_2 + V_3 + \cdots + V_n$$

Or, using Ohm's law

$$V = R_1 I + R_2 I + R_3 I + \cdots + R_n I = (R_1 + R_2 + R_3 + \cdots + R_n)I$$

Hence the entire system satisfies Ohm's law in the following way

$$V = R_T I$$

Where the total or equivalent resistance is

$$R_T = R_1 + R_2 + R_3 + \cdots + R_n \text{ (for resistors in series)} \tag{3.1}$$

This result allows us to replace a set of resistors connected in series by a single resistor whose resistance is given by (3.1).

EXAMPLE 3-1

Find the equivalent resistance seen at the two end terminals for the series circuit shown in Fig. 3-5.

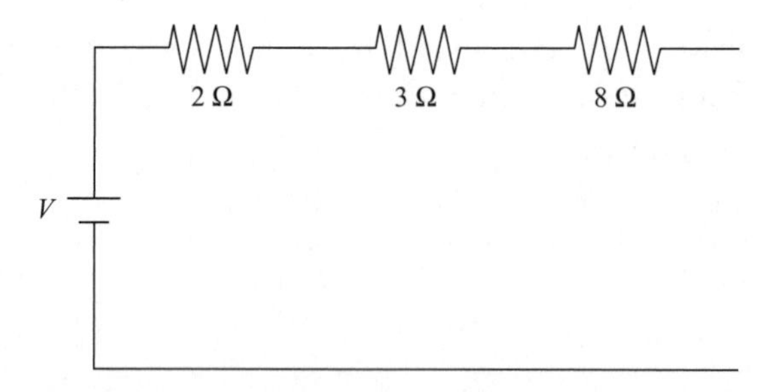

Fig. 3-5 A set of resistors connected in series solved in Example 3-1.

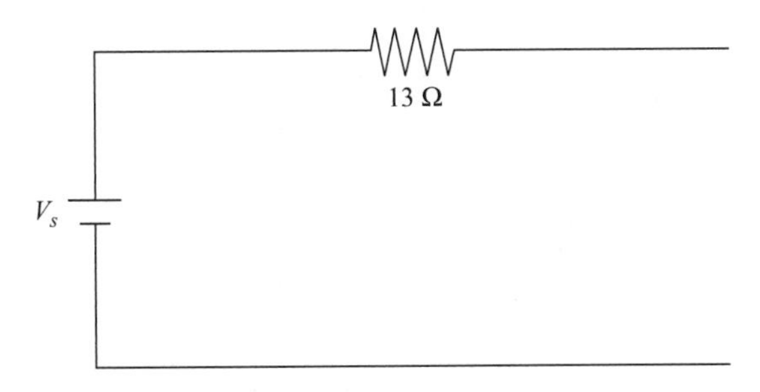

Fig. 3-6 The circuit shown here is equivalent to the circuit shown in Fig. 3-5.

SOLUTION

The total or equivalent resistance is found by adding up the values of resistance for each individual resistor. For the circuit shown in Fig. 3-5, we find

$$R_T = 2 + 3 + 8 = 13\ \Omega$$

Therefore, the circuit can be replaced by the equivalent circuit shown in Fig. 3-6.

The application of Ohm's law also allows us to simplify a set of resistors connected in parallel. Consider a set of resistors connected in parallel where the first resistor is connected across a voltage source, as shown in Fig. 3-7.

The current I that flows will be divided into currents I_1, I_2, etc., but the same voltage V_s is across each resistor (just apply KVL to each loop in the circuit to see this). With Ohm's law, the current that flows through the jth resistor is

$$I_j = \frac{V_s}{R_j}$$

This relation holds for each resistor in the circuit shown in Fig. 3-7. Applying KCL at the node where the first resistor is connected to the voltage source,

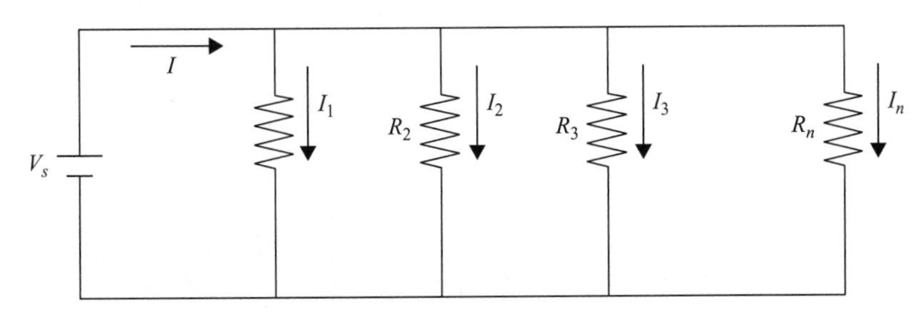

Fig. 3-7 A set of resistors connected in parallel.

we have

$$I = I_1 + I_2 + I_3 + \cdots + I_n$$

Using Ohm's law, we obtain an equivalent or total conductance G_T

$$\begin{aligned} I &= I_1 + I_2 + I_3 + \cdots + I_n \\ &= \frac{V_s}{R_1} + \frac{V_s}{R_2} + \frac{V_s}{R_3} + \cdots + \frac{V_s}{R_n} \\ &= V_s\left(\frac{1}{R_1} + \frac{1}{R_2} + \frac{1}{R_3} + \cdots + \frac{1}{R_n}\right) = V_s G_T \end{aligned}$$

Therefore, the equivalent or total resistance for a set of resistors connected in parallel is given by

$$\frac{1}{R_T} = \frac{1}{R_1} + \frac{1}{R_2} + \frac{1}{R_3} + \cdots + \frac{1}{R_n} \tag{3.2}$$

For the special case of two resistors connected in parallel

$$R_T = \frac{R_1 R_2}{R_1 + R_2} \tag{3.3}$$

EXAMPLE 3-2

Find the equivalent resistance for the circuit shown in Fig. 3-8, as seen by the voltage source.

SOLUTION

Using (3.2) we have

$$\frac{1}{R_T} = \frac{1}{R_1} + \frac{1}{R_2} = \frac{1}{3} + \frac{1}{2} = \frac{2}{6} + \frac{3}{6} = \frac{5}{6}$$

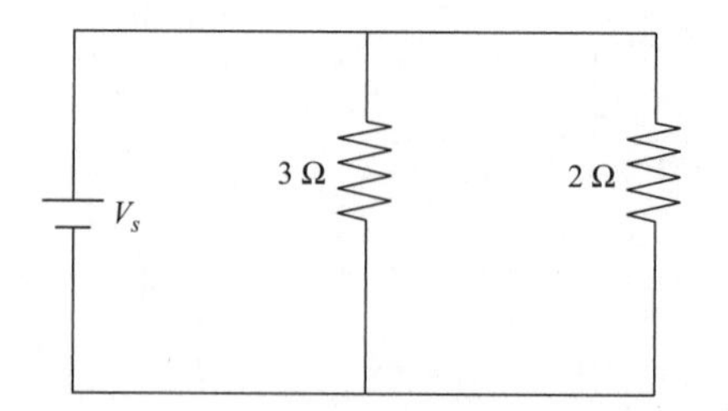

Fig. 3-8 The circuit used in Example 3-2.

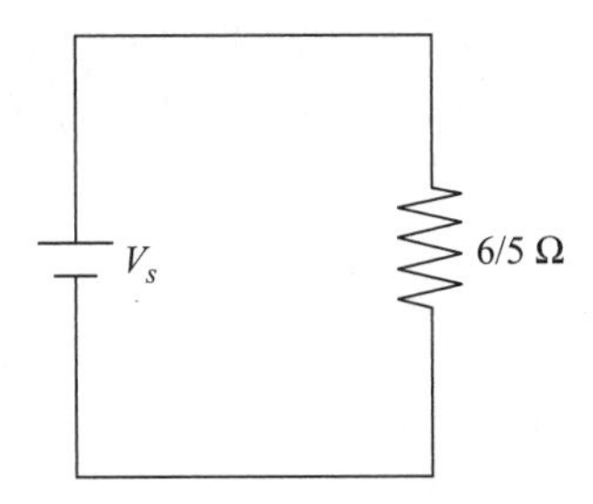

Fig. 3-9 Since the resistors in Fig. 3-8 are connected in parallel, the circuit can be replaced by this equivalent circuit.

Inverting gives us the equivalent or total resistance seen from the point of view of the voltage source

$$R_T = \frac{6}{5}\ \Omega$$

Hence the circuit can be replaced by the equivalent circuit shown in Fig. 3-9.

EXAMPLE 3-3

Simplify the circuit shown in Fig. 3-10.

SOLUTION

First we see immediately that the same current flows throw the 3 and 2 Ω resistors in the center of the network. They are in series and so

$$R_T = 3 + 2 = 5\ \Omega$$

The circuit can then be replaced by the one shown in Fig. 3-11.

It may not be immediately obvious, but the two 5 Ω resistors are in parallel. To see this recall that resistors that are in parallel have the same voltage across

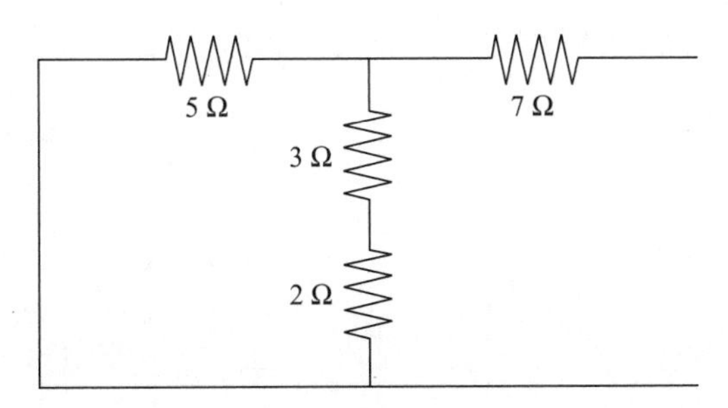

Fig. 3-10 The circuit used in Example 3-3 has resistors in series and in parallel.

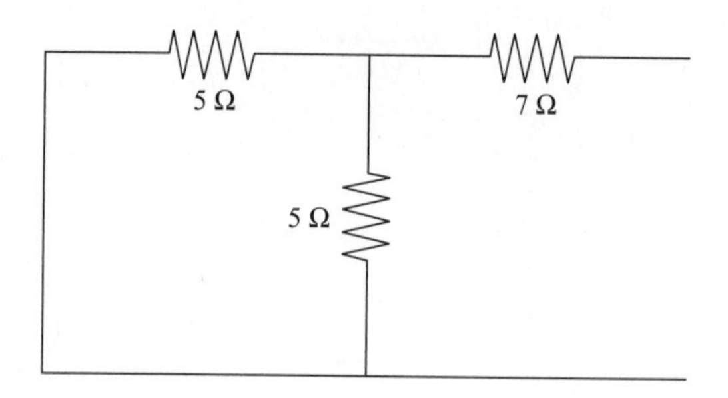

Fig. 3-11 The two resistors that were in series in Fig. 3-10 are replaced by the 5 Ω resistor.

them. Then apply KVL in a loop about the left-hand pane in Fig. 3-11. The equivalent resistance in this case is

$$\frac{1}{R_{\text{eq}}} = \frac{1}{5} + \frac{1}{5} = \frac{2}{5}$$

$$\Rightarrow R_{\text{eq}} = \frac{5}{2}\ \Omega$$

This means that we can replace the two 5 Ω resistors in Fig. 3-11 by a single 5/2 Ω resistor, which is in series with the 7 Ω resistor. This is shown in Fig. 3-12.

Now we are at a point where we can calculate the total resistance in the circuit. Since the remaining resistors are in series, we can replace them by a single resistor with resistance given by

$$R_T = 5/2 + 7 = \frac{19}{2}\ \Omega$$

The final circuit is shown in Fig. 3-13.

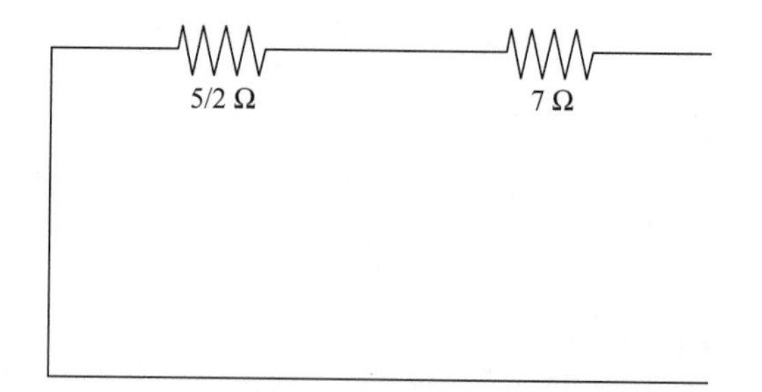

Fig. 3-12 We have reduced the circuit to two resistors in series.

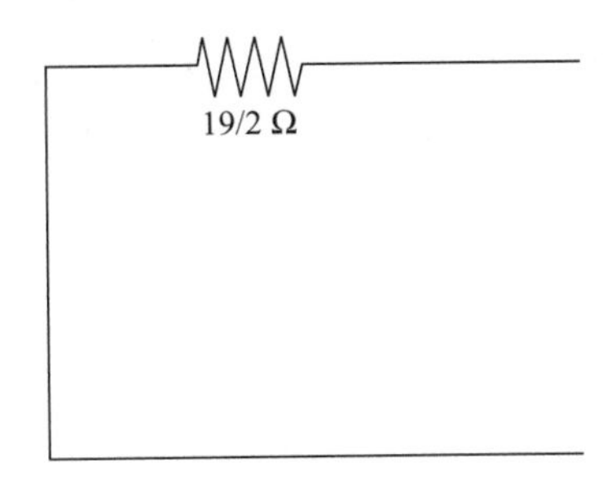

Fig. 3-13 This circuit is equivalent to the one shown in Fig. 3-10, as far as the two end terminals are concerned.

Back to Thevenin's Theorem

Now that we have seen how to combine resistors that are in series and in parallel, we can return to some examples that can be solved using Thevenin's theorem.

EXAMPLE 3-4

Find the Thevenin equivalent circuit connected to the left of the load resistor R_L as shown in Fig. 3-14.

SOLUTION

In this problem we want to replace the circuit to the left of the load resistor by a simpler Thevenin equivalent circuit. This circuit will consist of a single voltage source and a single resistor but will appear the same electrically to the load resistor as the circuit shown in Fig. 3-14. The first step is to disconnect the load and calculate the Thevenin equivalent voltage across the end terminals, as shown in Fig. 3-15.

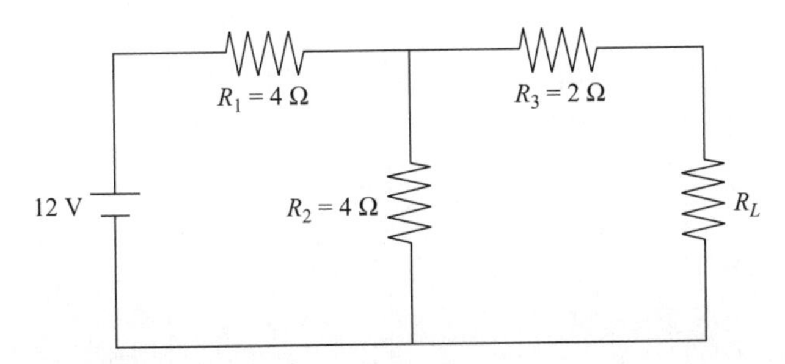

Fig. 3-14 In Example 3-4, we find the Thevenin equivalent circuit as seen by the load resistor R_L.

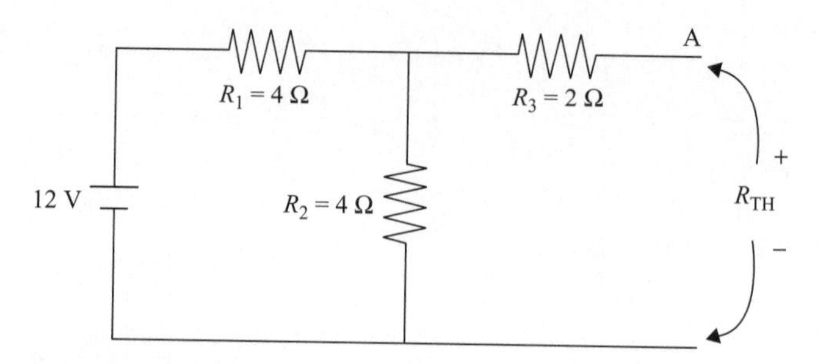

Fig. 3-15 The Thevenin equivalent voltage for Example 3-4.

We can find V_{TH} by noting first that the resistor R_3 does not affect the calculation. This is because one end of the resistor defines an open circuit; hence, no current is flowing through the resistor. By Ohm's law there is no voltage across it. This leaves the resistors R_1 and R_2, which define a voltage divider (they are in series). The voltage across R_1 is

$$V_1 = \frac{R_1}{R_1 + R_2} V_s = \frac{4}{4+4} 12 = 6\ \text{V}$$

Now we can apply KVL to a loop around the outside of the circuit

$$-V_s + V_1 + V_{TH} = 0$$
$$\Rightarrow V_{TH} = V_S - V_1 = 12 - 6 = 6\ \text{V}$$

Then we proceed to the next step in the Thevenin's theorem algorithm. We zero out all independent sources. In this case there is one independent voltage source, the 12 V battery on the left. We set $V_s = 0$ and replace it by a short circuit. This is shown in Fig. 3-16.

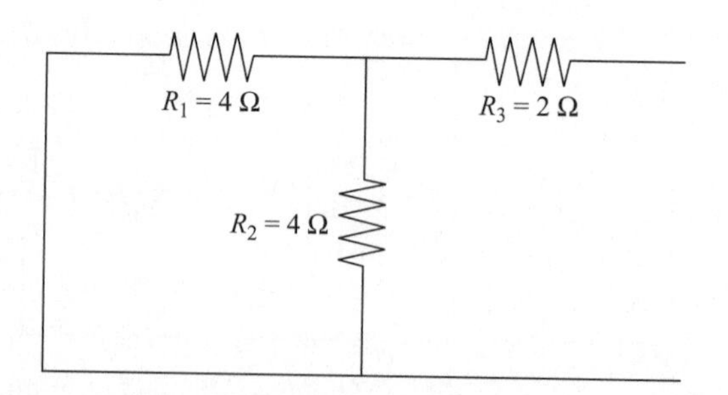

Fig. 3-16 Circuit with sources set to zero for Example 3-4.

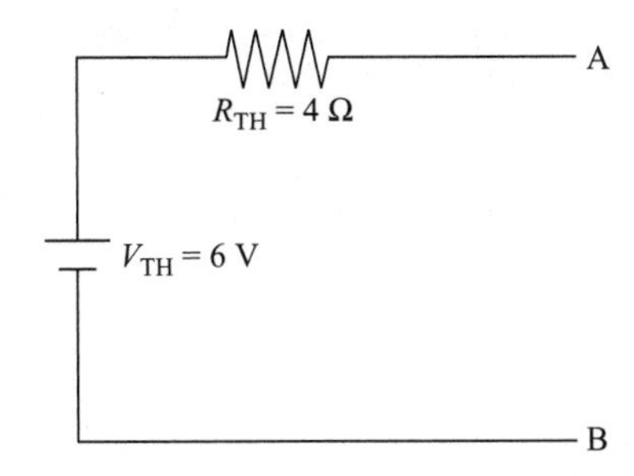

Fig. 3-17 The Thevenin equivalent circuit for Example 3-4.

Resistors R_1 and R_2 are in parallel, so we replace them by the resistor with equivalent resistance given by

$$\frac{1}{R_{eq}} = \frac{1}{R_1} + \frac{1}{R_2} = \frac{R_1 + R_2}{R_1 R_2}$$

$$\Rightarrow R_{eq} = \frac{R_1 R_2}{R_1 + R_2} = \frac{(4)(4)}{4+4} = 2\ \Omega$$

This set of parallel resistors is in series with the 2 Ω resistor. So the total or Thevenin equivalent resistance is

$$R_{TH} = 2 + 2 = 4\ \Omega$$

The Thevenin equivalent circuit is built by putting the Thevenin equivalent voltage in series with the Thevenin equivalent resistance. This is shown in Fig. 3-17.

EXAMPLE 3-5

In the circuit shown in Fig. 3-18, find the voltage across the load resistor if $R_1 = 2\ \Omega$, $R_2 = 3\ \Omega$, $R_3 = R_4 = 6\ \Omega$, the voltage source is $V_s = 15$ V, and the current is $I_L = 3$ A.

SOLUTION

The first step is to remove the load resistor and calculate the Thevenin equivalent voltage across the resulting open circuit. This is shown in Fig. 3-19.

We take the positive reference for the voltage across each resistor to be at the top of the resistor. This looks complicated, but it really isn't. If we ignore the two inside resistors R_1 and R_3 for the moment, it should be easy to deduce that resistors R_2 and R_4 are in series. Then we can apply KVL around the external

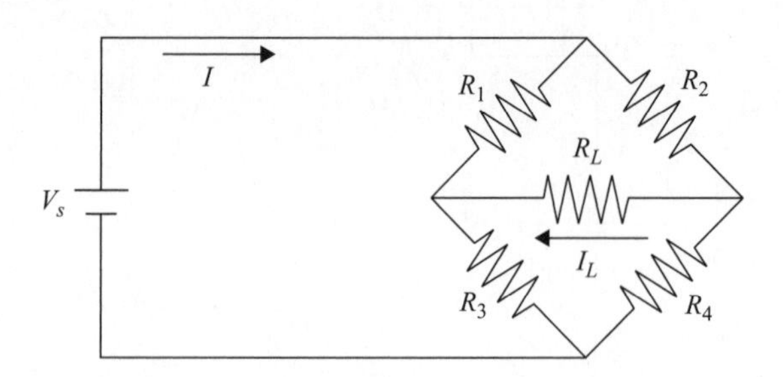

Fig. 3-18 The circuit studied in Example 3-5.

loop. The circuit for which we apply KVL looks like the simple network shown in Fig. 3-20.

So we have two voltage dividers. The voltage across each resistor is (using the values given in the problem statement)

$$V_2 = \frac{R_2}{R_2 + R_4} V_s = \frac{3}{3+6} 15 = 5 \text{ V}$$

$$V_4 = \frac{R_4}{R_2 + R_4} V_s = \frac{6}{3+6} 15 = 10 \text{ V}$$

A similar procedure can be applied to the resistors R_1 and R_3, giving voltage dividers

$$V_1 = \frac{R_1}{R_1 + R_3} V_s = \frac{2}{2+6} 15 = 3.75 \text{ V}$$

$$V_3 = \frac{R_3}{R_1 + R_3} V_s = \frac{6}{2+6} 15 = 11.25 \text{ V}$$

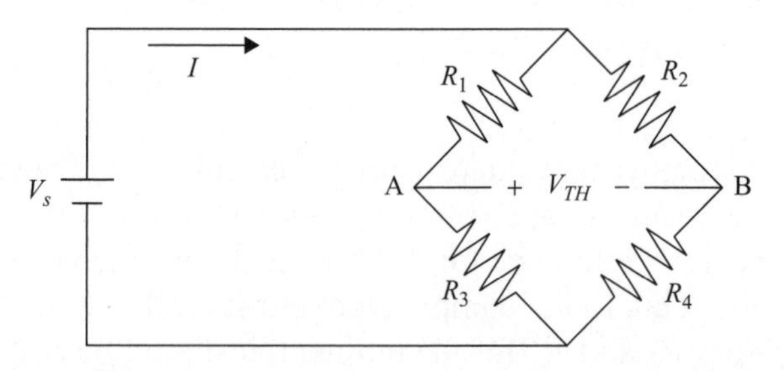

Fig. 3-19 The first step in Example 3-5 is to calculate the Thevenin equivalent voltage.

Fig. 3-20 The two resistors on the right-hand edge of Fig. 3-19 are in series.

Notice that in each case the conservation of energy is satisfied; that is, the sum of the voltages across R_2 and R_4 and the sum of the voltages across R_1 and R_3 are equal to the value of the voltage source

$$V_2 + V_4 = 5 + 10 = 15 \text{ V}$$

$$V_1 + V_3 = 3.75 + 11.25 = 15 \text{ V}$$

Now we know the voltage across every resistor in Fig. 3-19 and can find the Thevenin equivalent voltage by taking any loop we like and applying KVL. Taking the top loop around V_{TH}, R_1, and R_2 we have

$$-V_{TH} - V_1 + V_2 = 0$$

$$\Rightarrow V_{TH} = V_2 - V_1 = 5 - 3.75 = 1.25 \text{ V}$$

Let's find the Thevenin equivalent resistance. First we set the voltage source equal to zero and replace it by a short circuit. Then we have the circuit shown in Fig. 3-21.

The combinations $R_1 - R_3$ and $R_2 - R_4$ are in parallel. So the circuit shown in Fig. 3-21 is equivalent to the circuit shown in Fig. 3-22.

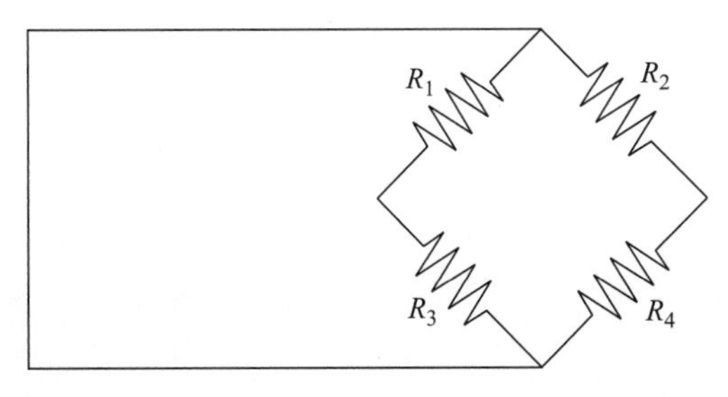

Fig. 3-21 The circuit used to find the equivalent resistance for Example 3-5.

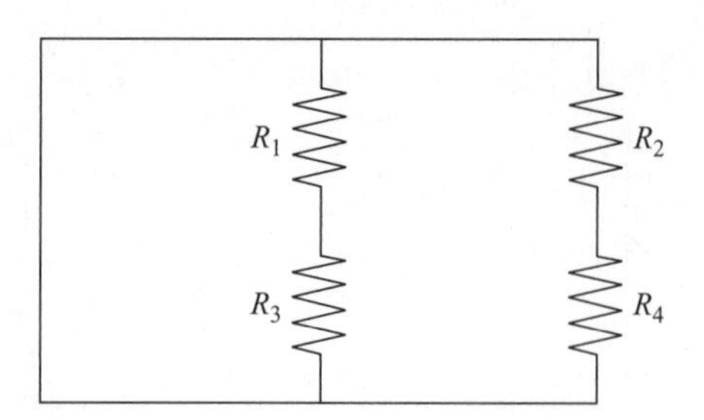

Fig. 3-22 A clear way to view the circuit in Fig. 3-21.

The equivalent resistance is

$$\frac{1}{R_{TH}} = \frac{1}{R_1 + R_3} + \frac{1}{R_2 + R_4} \Rightarrow R_{TH} = \frac{(R_1 + R_3)(R_2 + R_4)}{R_1 + R_3 + R_2 + R_4}$$
$$= \frac{(2+6)(3+6)}{2+6+3+6} = 4.2\ \Omega$$

Remember, since each combination shown in Fig. 3-22 is in series, we add them together to get the equivalent resistance.

We can then replace the complex circuit with the resistors connected in a diamond shape (Fig. 3-19) by the Thevenin equivalent circuit shown in Fig. 3-23, which is much simpler.

Now we can find the voltage in the load resistor. Reconnecting it, we need to solve the circuit shown in Fig. 3-24.

Applying KVL to the loop, we have

$$-V_{TH} + R_{TH}I + V_L = 0$$
$$\Rightarrow V_L = V_{TH} - R_{TH}I = 1.25 - (3)(4.2) = -11.4\ \text{V}$$

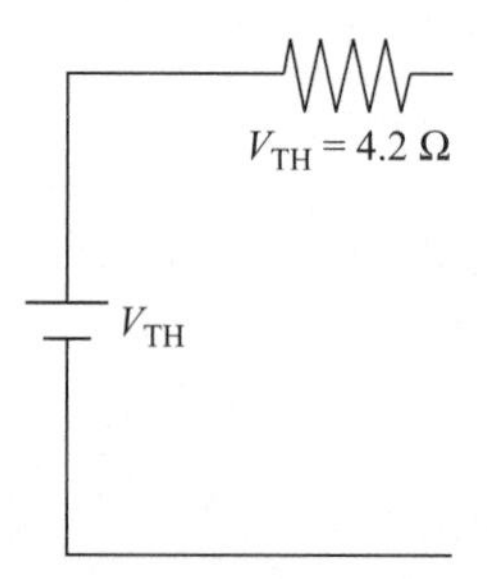

Fig. 3-23 The Thevenin equivalent circuit for Example 3-5.

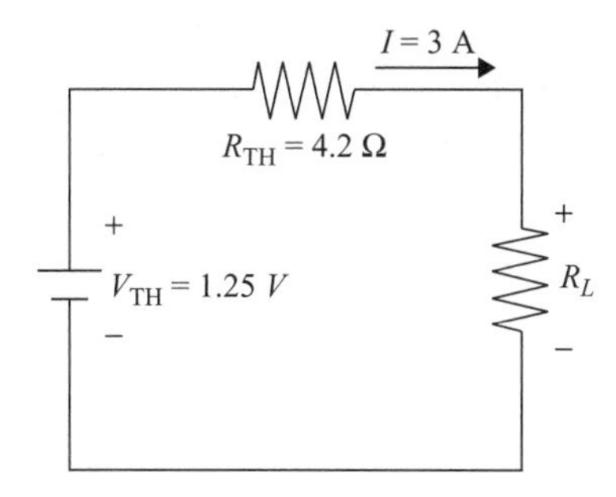

Fig. 3-24 The load resistor connected to the Thevenin equivalent circuit.

All the minus sign means is that the polarity of the load voltage is *opposite* to that we chose in Fig. 3-24.

EXAMPLE 3-6

Find the power in the load resistor for the circuit shown in Fig. 3-25 by using Thevenin's theorem. The load resistor is 5 Ω.

SOLUTION

The first step is to detach the load resistor and calculate the Thevenin equivalent voltage across the resulting open circuit. Note that the 1 and 2 Ω resistors are in parallel. They can be replaced with the equivalent resistance

$$\frac{1}{R_{\text{eq}}} = \frac{1}{2} + 1 = \frac{3}{2}$$

$$\Rightarrow R_{\text{eq}} = \frac{2}{3}\ \Omega$$

Hence we can proceed with the circuit shown in Fig. 3-26.

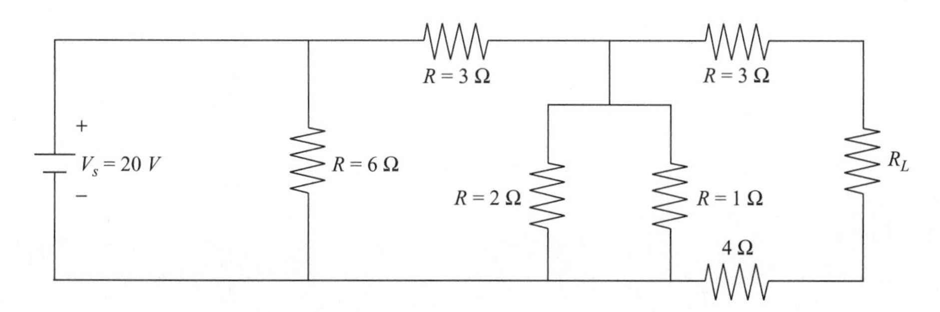

Fig. 3-25 The circuit studied in Example 3-6.

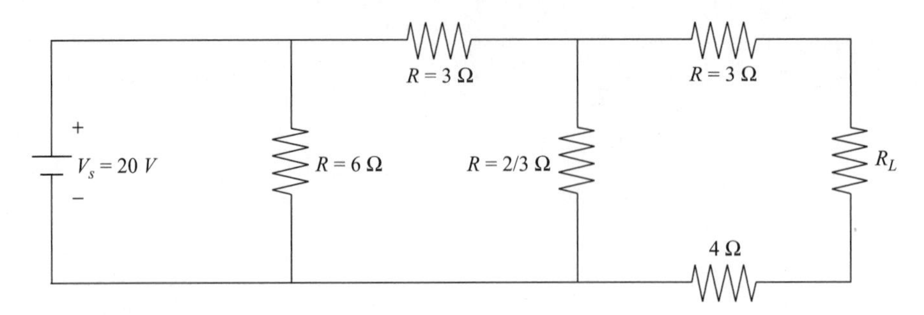

Fig. 3-26 The equivalent resistance found by combining the 2 and 1 Ω resistors that were in parallel in Fig. 3-25.

Notice we can take a KVL loop around the voltage source, the 3 Ω resistor, and the 2/3 Ω resistor. Hence the voltages across each can be found by using voltage dividers

$$V_3 = \frac{3}{3 + 2/3} 20 = 16.4 \text{ V}$$

V_{TH}

$$V_{2/3} = \frac{2/3}{3 + 2/3} 20 = 3.6 \text{ V}$$

Removing the load resistor, the Thevenin equivalent voltage will be the voltage across the open circuit terminals at the 3 and 4 Ω resistors on the right-hand side of Fig. 3-27.

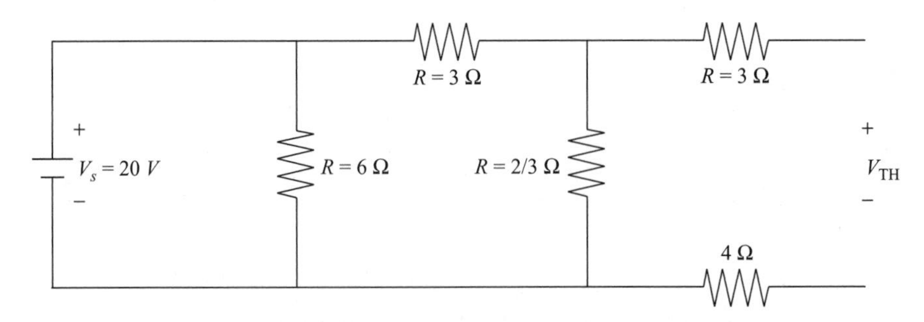

Fig. 3-27 We can find the Thevenin equivalent voltage by applying KVL to this circuit.

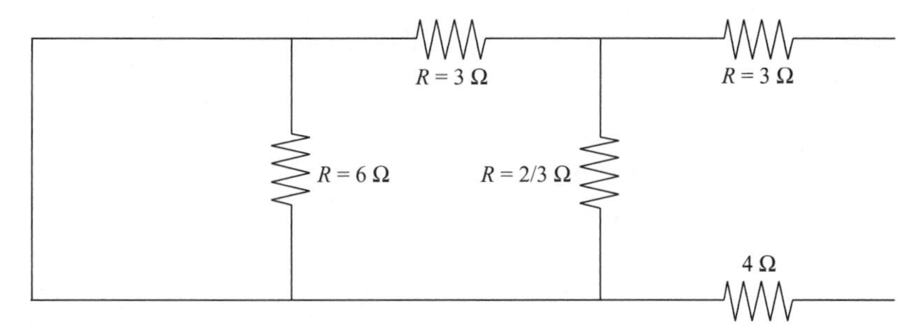

Fig. 3-28 We start to find the Thevenin equivalent resistance with this circuit.

In fact, we can just take KVL around the right-hand loop

$$-V_{2/3} + V_{\text{TH}} = 0$$
$$\Rightarrow V_{\text{TH}} = V_{2/3} = 3.6 \text{ V}$$

Now let's zero out the voltage source and replace it by a short circuit. This is shown in Fig. 3-28.

Now the 6 Ω resistor and the short circuit are in parallel, and there is no voltage across a short circuit. Hence the voltage across the 6 Ω resistor is also zero. It's the same as saying that the resistor isn't there at all, so we can replace it with the circuit shown in Fig. 3-29.

The 2/3 and 3 Ω resistors in Fig. 3-29 are in parallel. The equivalent resistance is

$$\frac{1}{R_{\text{eq}}} = \frac{1}{3} + \frac{3}{2} = \frac{11}{6}, \Rightarrow R_{\text{eq}} = \frac{6}{11} \ \Omega$$

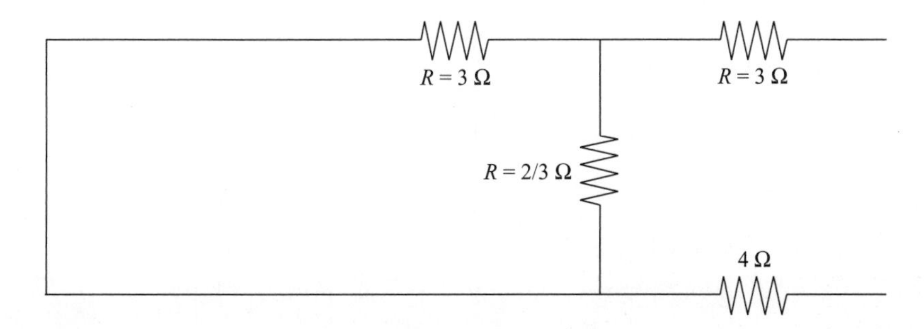

Fig. 3-29 Removing the resistor which becomes irrelevant because of the short in Fig. 3-28 leaves this circuit.

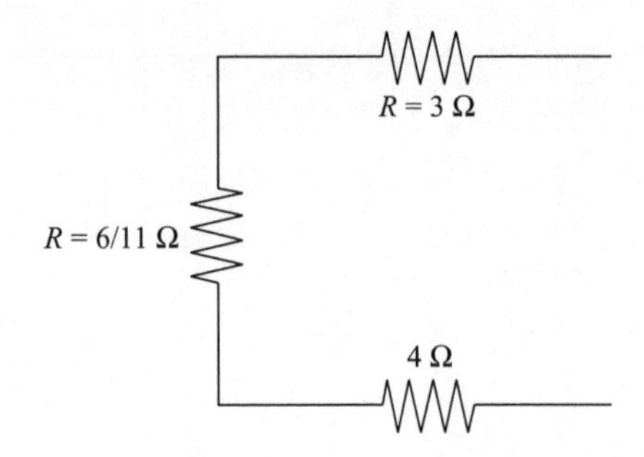

Fig. 3-30 A simplification of the circuit shown in Fig. 3-29. The Thevenin equivalent resistance can then be found by adding up these resistances in series.

We can use this to simplify the circuit in Fig. 3-29 and replace it with the circuit shown in Fig. 3-30.

The remaining resistors are in series, so we just add them up to get the Thevenin equivalent resistance

$$R_{TH} = \frac{6}{11} + 3 + 4 = \frac{83}{11}\ \Omega$$

The Thevenin equivalent circuit is shown, with the load resistor attached, in Fig. 3-31.

The current flowing through the resistor is

$$I = \frac{V_{TH}}{R_{TH} + R_L} = \frac{3.6}{83/11 + 5} = 0.3\ \text{A}$$

The power is

$$P = VI = RI^2 = (5)(0.3)^2 = 0.41\ \text{A}$$

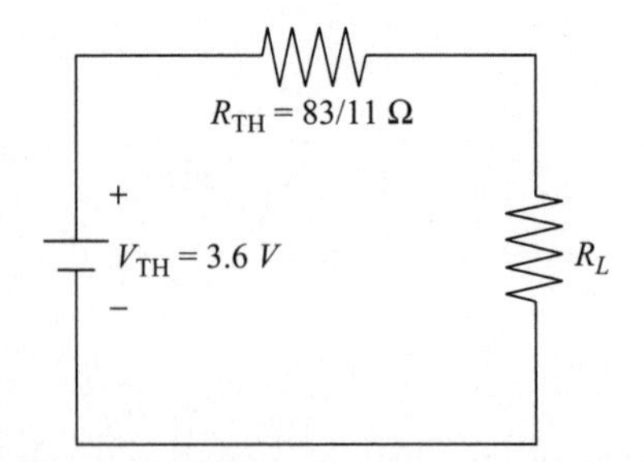

Fig. 3-31 The Thevenin equivalent circuit attached to the load resistor in Example 3-6.

Thevenin's Theorem Using the Karni Method

When I took circuit analysis at the University of New Mexico in Albuquerque, I was fortunate to have Shlomo Karni as my professor. Dr. Karni encouraged the students to work hard in his courses, assigning homework every day, giving him a bad reputation among the students. Of course later we saw the benefits of having been worked so hard.

In his circuits course Dr. Karni introduced a clever and simple way to calculate the Thevenin resistance and voltage. In the course I took from him, he told us we were the first class to ever learn the method. To my knowledge Dr. Karni is the originator of this technique, so we will refer to this way of applying Thevenin's theorem as the *Karni method.*

The basic idea behind the Karni method is the following. We again consider an outside observer or network connected to some circuit at terminals A and B as shown in Fig. 3-1. Instead of analyzing the network by considering an open circuit at A–B, we do the following:

- Add a *current source* I_o to terminals A–B.
- Calculate the voltage v_o across this current source.

The voltage across the applied current source will be expressed in the form

$$v_o = R_{\mathrm{TH}} I_o + V_{\mathrm{TH}} \tag{3.4}$$

This will allow us to just read off the Thevenin equivalent resistance and voltage.

Note that the actual value of the current source and voltage are irrelevant, so we leave them as symbolic currents and voltages. The Karni method is best illustrated with examples. Let's look at two scenarios.

EXAMPLE 3-7

Find the Thevenin equivalent of the circuit shown in Fig. 3-32.

SOLUTION

The first step is to add a current source to the open-circuit terminals A-B. This is shown in Fig. 3-33.

Now we apply KCL and KVL to the circuit. Applying KVL to the outside loop as shown in Fig. 3-33, we have

$$-100 + 5I_1 + 6I_o + v_o = 0$$
$$\Rightarrow v_o = 100 - 5I_1 - 6I_o$$

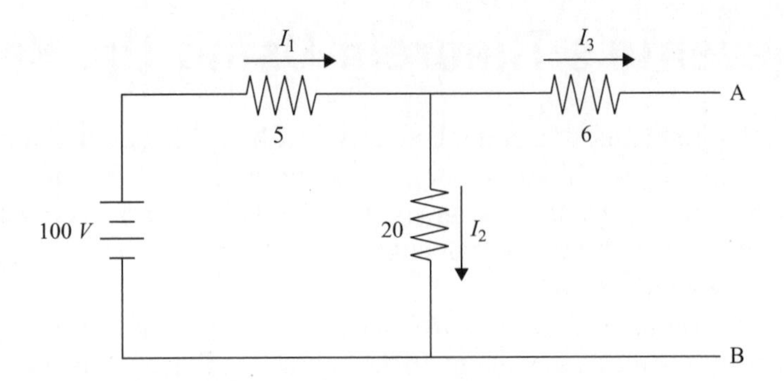

Fig. 3-32 We solve this circuit by using the Karni method in Example 3-7.

We can apply KCL (+ for currents leaving) to obtain an expression for the unknown current I_1. Clearly at the top node we have

$$I_1 = I_o + I_2 \tag{3.5}$$

Now let's apply KVL to the left pane in Fig. 3-33. Using Ohm's law, we obtain

$$5I_1 + 20I_2 = 100$$

Or, solving for I_2

$$I_2 = 5 - (1/4)\, I_1 \tag{3.6}$$

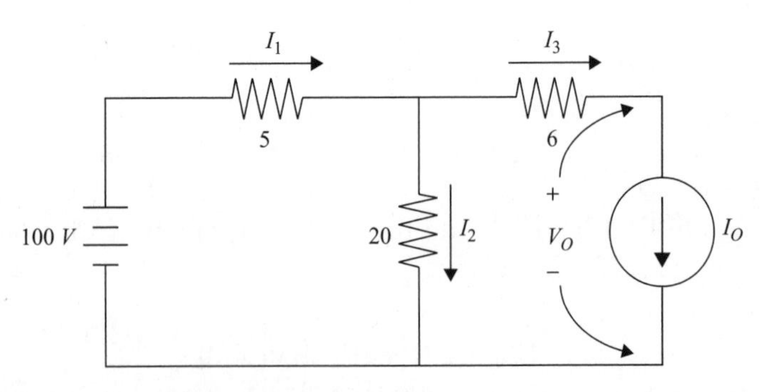

Fig. 3-33 The first step in applying the Karni method is to add a current source.

Now we can eliminate I_2 in (3.5). This gives

$$I_1 = I_o + I_2 = I_o + 5 - \frac{1}{4} I_1$$

$$\Rightarrow I_1 = 4 + \frac{4}{5} I_o$$

Let's use this in our expression for v_o. Remember we found that $v_o = 100 - 5I_1 - 6I_o$. Hence we find that

$$v_o = 100 - 20 - 4I_o - 6I_o = 80 - 10I_o$$

Comparison with (3.4) allows us to read off the Thevenin equivalent resistance and voltage, where we find that

$$R_{\text{TH}} = 10\ \Omega$$
$$V_{\text{TH}} = 80\ \text{V}$$

EXAMPLE 3-8

Find the Thevenin equivalent circuit to the network shown in Fig. 3-34 as seen at the sinusoidal voltage source. Then calculate the current in the voltage source.

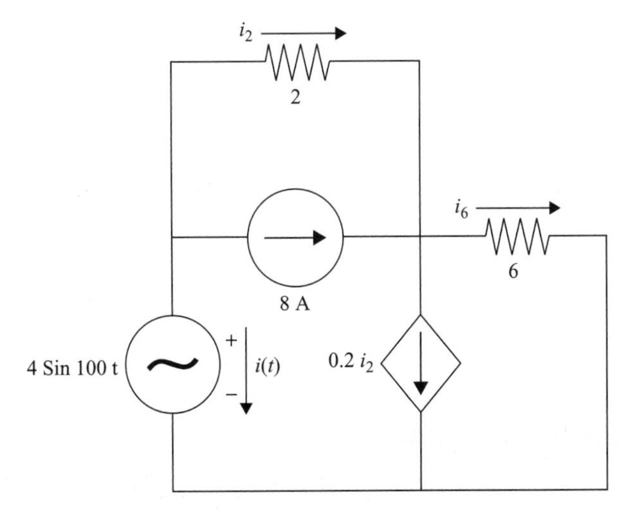

Fig. 3-34 The circuit solved with the Karni method in Example 3-8.

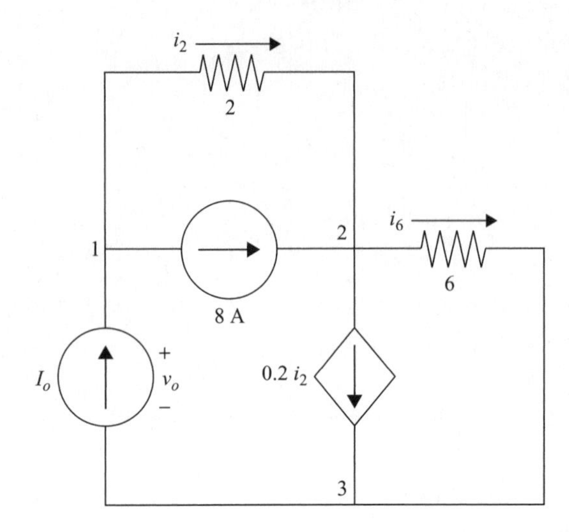

Fig. 3-35 The circuit obtained from Fig. 3-34 by using the Karni method.

SOLUTION

If we want to find the Thevenin equivalent network as seen by a given component using the Karni method, then we remove that component and replace it with a current source that we label I_o. This is shown in Fig. 3-35.

Now we can solve the circuit by using ordinary techniques, i.e., we apply KCL and KVL. First let's apply KCL at node 1 (as seen in Fig. 3-35), taking + for currents leaving the node. We have

$$\begin{aligned} &-I_o + i_2 + 8 = 0 \\ &\Rightarrow i_2 = I_o - 8 \end{aligned} \tag{3.7}$$

The goal is to write all unknown quantities in terms of the current source I_o, which is why we solved for i_2 in this case. Now let's apply KCL again, this time at node 2 as shown in Fig. 3-35. This time we have

$$\begin{aligned} &-8 - i_2 + 0.2\, i_2 + i_6 = 0 \\ &\Rightarrow i_6 = 8 + 0.8\, i_2 \end{aligned} \tag{3.8}$$

Using (3.7), we can rewrite this as

$$i_6 = 8 + 0.8\,(I_o - 8) = 8 + 0.8 I_o - 6.4 \tag{3.9}$$

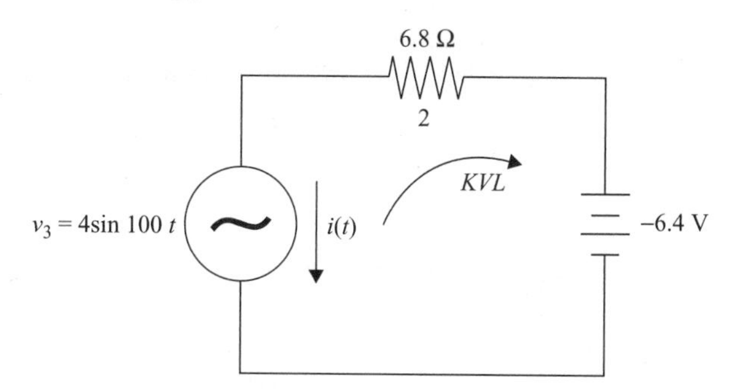

Fig. 3-36 The Thevenin equivalent ciruit to the one shown in Fig. 3-34, as seen by the voltage source.

giving us the second unknown current in terms of the current source I_o. Now let's apply KVL. For our loop, we take the outside loop around the circuit, giving

$$-v_o + 2\,i_2 + 6\,i_6 = 0$$

Using (3.7) and (3.9) we can rewrite this as

$$v_o = 6.8 I_o - 6.4 \tag{3.10}$$

Take a look back at (3.4). This equation allows us to read off the Thevenin equivalent resistance and voltage

$$R_{\mathrm{TH}} = 6.8\ \Omega, \quad V_{\mathrm{TH}} = -6.4\ \mathrm{V} \tag{3.11}$$

The Thevenin equivalent circuit is shown below in Fig. 3-36. Remember, the circuit we were seeking was the circuit *as seen by the voltage source.*

With this much simpler network in hand, we can easily find the current that passes through the voltage source. We label the current $i(t)$ and apply KVL to the circuit. We find

$$4 \sin 100t - 6.8\,i(t) - 6.4 = 0$$

Hence

$$i(t) = \frac{-6.4 + 4 \sin 100t}{6.8} = -0.94 + 0.6 \sin 100t \quad [\mathrm{A}]$$

The Karni method is easier to use in many cases than the standard application of Thevenin's theorem. The few times you use it, apply both methods and compare your answers to ensure you understand how to apply it correctly, or check it against problems solved with Thevenin's theorem in your textbook.

Norton's Theorem and Norton Equivalent Circuits

Another method that can be used to analyze circuits is known as *Norton's Theorem.* In this case, we wish to find the Norton current I_N and the Norton resistance R_N for a given circuit consisting of sources and resistors that may be arranged in some complicated fashion. This really isn't all that different from Thevenin's theorem. In this case, we will replace the given network with one that has a current source and resistor arranged in parallel.

The Norton resistance is nothing other than the Thevenin equivalent resistance

$$R_N = R_{\text{TH}} \tag{3.12}$$

The Norton current is found using Ohm's law as applied to the Thevenin voltage and resistance. That is

$$I_N = \frac{V_{\text{TH}}}{R_{\text{TH}}} \tag{3.13}$$

So you can see that there really isn't much new here. We first find the Thevenin equivalent circuit and then determine the Norton current to obtain the value of the current source. Then we have the Norton equivalent circuit as shown in Fig. 3-37.

This theorem comes in handy if for some reason you need a circuit with the components in parallel and would prefer a current source instead of a voltage source. Since there really isn't anything new here, we will take a quick look at the method in a single example.

EXAMPLE 3-9

Find the Norton equivalent circuit as seen by the load resistor in Fig. 3-38.

SOLUTION

The first step is to disconnect the load resistor R_L and find the open-circuit voltage across terminals A–B. For the circuit shown in Fig. 3-38, with the load

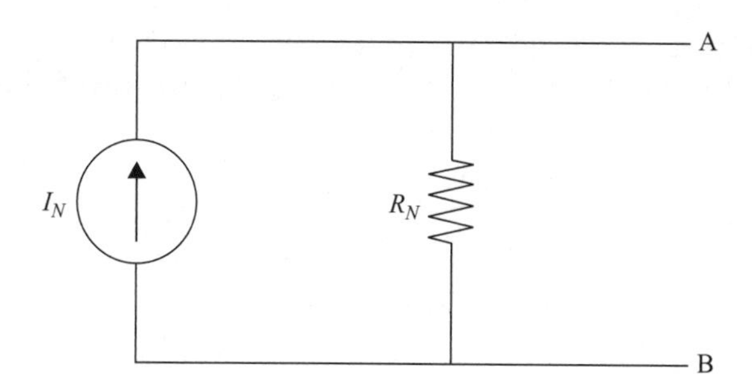

Fig. 3-37 The Norton equivalent circuit as seen at terminals A and B.

resistor replaced by an open circuit, the voltage across A-B is given by a voltage divider

$$V_{TH} = \frac{V_S R_2}{R_1 + R_2}$$

Now, resistors R_1 and R_2 are in parallel, so can be replaced by

$$\frac{R_1 R_2}{R_1 + R_2}$$

This resistance is in series with R_3, so the total resistance is

$$R_{TH} = \frac{R_1 R_2}{R_1 + R_2} + R_3$$

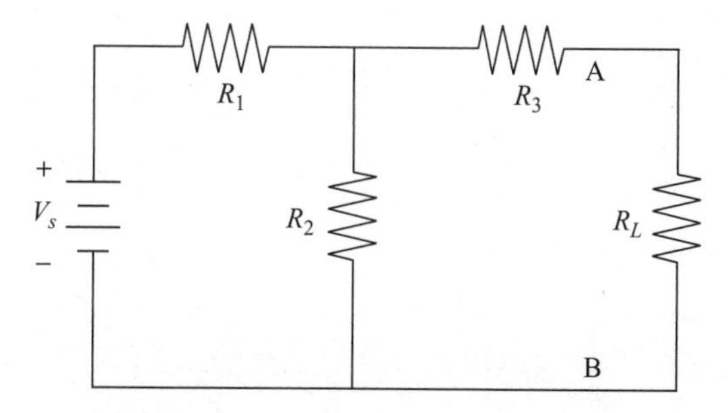

Fig. 3-38 In Example 3-9, we want to find the Norton equivalent circuit as seen at terminals A and B by the load resistor.

This is the same as the Norton resistance R_N, so the only piece of work remaining is to find the Norton current by using Ohm's law. We obtain

$$I_N = \frac{V_{\text{TH}}}{R_{\text{TH}}} = \frac{\dfrac{V_S R_2}{R_1 + R_2}}{\dfrac{R_1 R_2}{R_1 + R_2} + R_3} = \frac{V_S R_2}{R_1 R_2 + R_1 R_3 + R_2 R_3}$$

The Norton equivalent circuit is then obtained by arranging the current source and resistance in parallel as shown in Fig. 3-37.

Summary

Thevenin's theorem is a powerful technique that can be used to simplify a complicated circuit consisting of sources and resistors that can be arranged in arbitrary parallel and series connections. A Thenvenin equivalent circuit is built consisting of a single voltage source in series with a single resistor. To apply Thevenin's theorem using the Karni method, attach an arbitrary current source at the location where you have the open circuit in the standard application of Thevenin's theorem. The Karni method greatly simplifies calculations. Finally, to replace a circuit with a single current source and resistor arranged in parallel, apply Norton's theorem.

Quiz

1. What is the equivalent resistance for the two resistors in the network shown in Fig. 3-39?

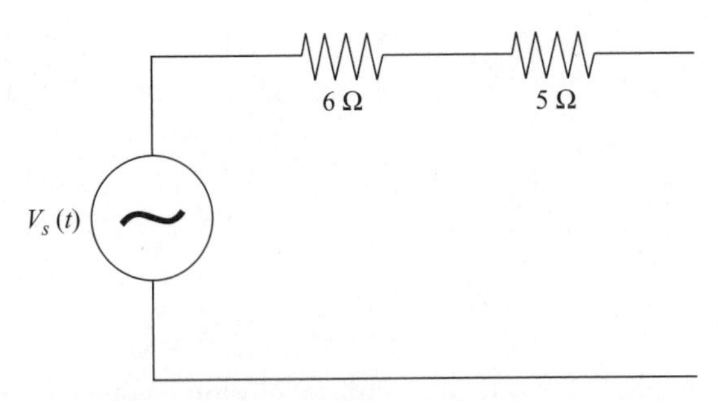

Fig. 3-39 Circuit for Problem 1.

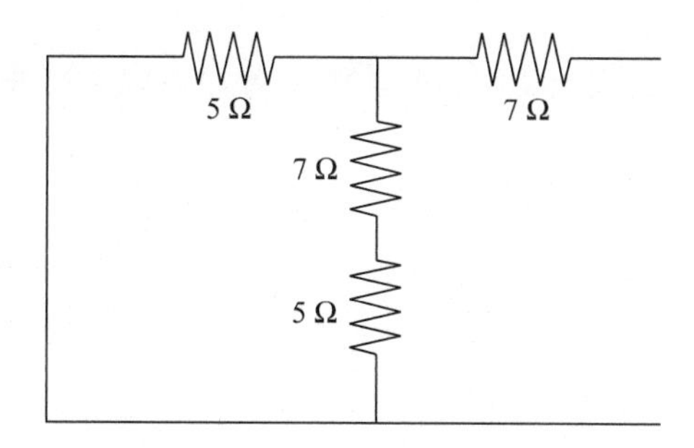

Fig. 3-40 Network for Problem 3.

2. What is the equivalent resistance if a 6 Ω resistor is in parallel with a 4 Ω resistor?
3. Find the equivalent resistance for the circuit in Fig. 3-40.
4. Find the Thevenin equivalent voltage and resistance for the circuit shown in Fig. 3-41. Do this using Thevenin's theorem, and then show that you get the same answer using the Karni method.
5. Find the current $i(t)$ in Fig. 3-41.

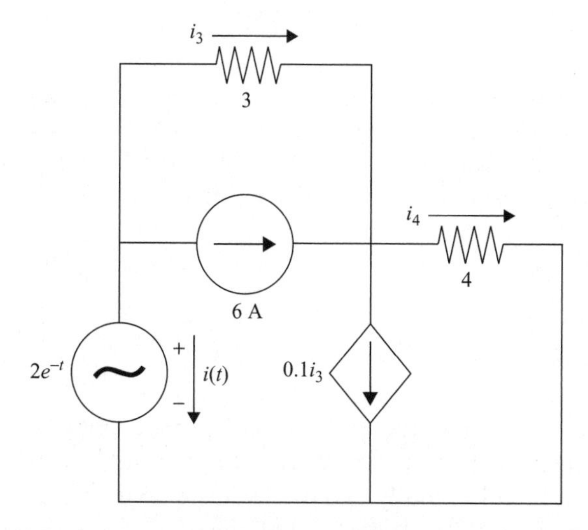

Fig. 3-41 Circuit for Problems 4 and 5. Borrowed from a circuit analysis exam given by Shlomo Karni in 1990.

CHAPTER 4

Network Theorems

We have already seen two important network theorems—Thevenin's and Norton's theorems. In this chapter we will introduce other theorems that can be used to simplify network analysis.

Superposition

Consider a circuit that contains multiple voltage and current sources. If all of the elements in the circuit are linear, we can simplify analysis by considering the effect of each source individually and then adding up the results. How does this work? A given component will have a voltage across it and a current through it due to each source. We calculate those voltages and currents considering each source individually. Then we add up all the currents and all the voltages to get the total current and voltage for the component. This is the essence of the *superposition theorem.* Let's quantify this. Let a circuit contain a set of n sources where the ith source is denoted s_i. Now suppose that the response of the circuit to s_i alone is $a_i s_i$, where a_i is a constant. Then the total response

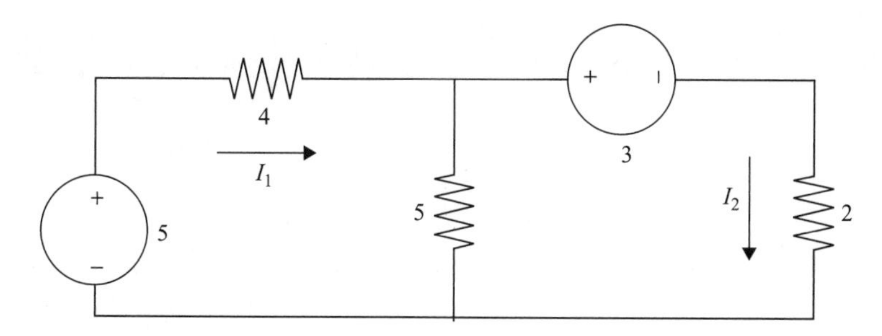

Fig. 4-1 We can apply the superposition theorem to this linear circuit, because it contains two sources.

of the circuit to all of the sources is found by adding up the individual responses

$$r = a_1 s_1 + a_2 s_2 + \cdots + a_n s_n \tag{4.1}$$

Let's see how to apply the superposition theorem to the circuit shown in Fig. 4-1.

EXAMPLE 4-1

Use the superposition theorem to find the current through the 4 Ω resistor in Fig. 4-1.

SOLUTION

First let's solve the circuit by using ordinary techniques. We begin by applying KVL to the left pane in Fig. 4-1. We find

$$-5 + 4I_1 + 5(I_1 - I_2) = 0$$

Collecting and rearranging terms gives

$$9I_1 - 5I_2 = 5$$

Now we apply KVL to the right pane in Fig. 4-1. We have

$$3 + 2I_2 + 5(I_2 - I_1) = 0$$

Collecting terms we obtain

$$-5I_1 + 7I_2 = -3$$

To find the two unknown currents we arrange the terms as the following system of equations

$$\left[\begin{array}{cc|c} 9 & -5 & 5 \\ -5 & 7 & -3 \end{array}\right]$$

The matrix on the left contains the coefficients of each current (resistances) while the column vector on the right is due to the sources. We use Cramer's rule to determine the value of the two unknown currents. For the first current we have

$$I_1 = \frac{\begin{vmatrix} 5 & -5 \\ -3 & 7 \end{vmatrix}}{\begin{vmatrix} 9 & -5 \\ -5 & 7 \end{vmatrix}} = \frac{35 - 15}{63 - 25} = \frac{20}{38} \text{ A}$$

This is the unknown current through the 4 Ω resistor that we will find using superposition in a moment. The other current is found to be

$$I_2 = \frac{\begin{vmatrix} 9 & 5 \\ -5 & -3 \end{vmatrix}}{\begin{vmatrix} 9 & -5 \\ -5 & 7 \end{vmatrix}} = \frac{-27 + 25}{63 - 25} = -\frac{8}{38} \text{ A}$$

Now let's solve for the first current using superposition. We will do this by replacing each voltage source by a short circuit in turn. We begin by leaving the 5 V source intact and setting the 3 V source to zero. This results in the circuit shown in Fig. 4-2.

We will illustrate the method by going through the same process of applying KVL. In the first pane, the equation is

$$-5 + 4I_1 + 5(I_1 - I_2) = 0$$

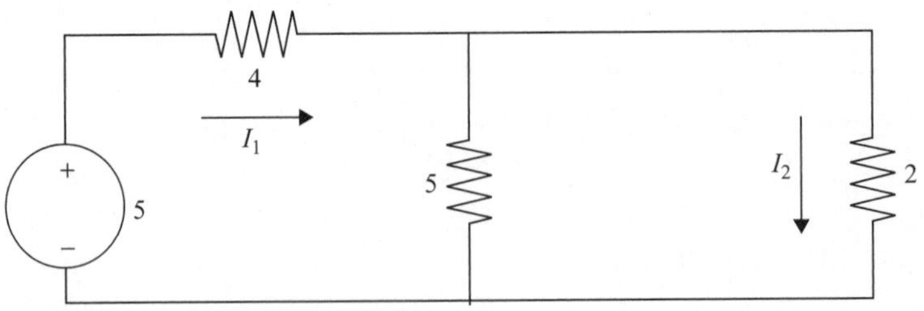

Fig. 4-2 Step one in superposition is to zero out the 3 V source.

which is the same result we had before—not surprising since the source in this pane is intact

$$9I_1 - 5I_2 = 5$$

For the right pane, the equation is

$$2I_2 + 5(I_2 - I_1) = 0$$

Collecting terms we have

$$-5I_1 + 7I_2 = 0$$

Now we can use Cramer's rule to find the current. But let's add a prime to the symbol used to label the current, because this is an intermediate value in our calculations

$$I_1' = \frac{\begin{vmatrix} 5 & -5 \\ 0 & 7 \end{vmatrix}}{\begin{vmatrix} 9 & -5 \\ -5 & 7 \end{vmatrix}} = \frac{35}{63 - 25} = \frac{35}{38}\text{ A}$$

Now we do the procedure again, this time setting the 5 V source to zero and leaving the 3 V source intact. The result is the circuit shown in Fig. 4-3.

Applying KVL to the left pane gives

$$4I_1 + 5(I_1 - I_2) = 0$$

Hence

$$9I_1 - 5I_2 = 0$$

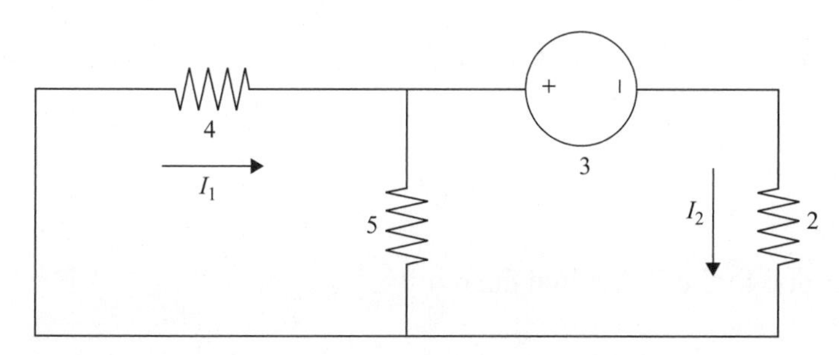

Fig. 4-3 We zero out the 5 V source and consider the effect of the 3 V source alone.

KVL applied to the right pane results in the same equation we originally obtained, namely

$$-5I_1 + 7I_2 = -3$$

Now we apply Cramer's rule again to get the second intermediate current due to the 3 V source alone. We denote this intermediate current with a double prime

$$I_1'' = \frac{\begin{vmatrix} 0 & -5 \\ -3 & 7 \end{vmatrix}}{\begin{vmatrix} 9 & -5 \\ -5 & 7 \end{vmatrix}} = -\frac{15}{38}\text{ A}$$

The superposition theorem tells us that the total current is due to the sums of the currents due to each individual source alone. That is

$$I_1 = I_1' + I_1'' = \frac{35}{38} - \frac{15}{38} = \frac{20}{38}\text{ A}$$

Some notes about superposition:

- Dependent sources cannot be set to zero when analyzing a circuit by using superposition.
- While superposition *can* simplify analysis, it doesn't always do so, as the previous example showed.
- Superposition cannot be used to perform power calculations, because power is either the product of voltage and current or the square of either one, so it's nonlinear. Superposition only works in the linear case.

Let's think about the previous example. Given the current the power for a resistor is

$$P = VI = RI^2$$

The power due to the individual currents is

$$P_1 = R(I_1')^2, \quad P_2 = R(I_1'')^2$$

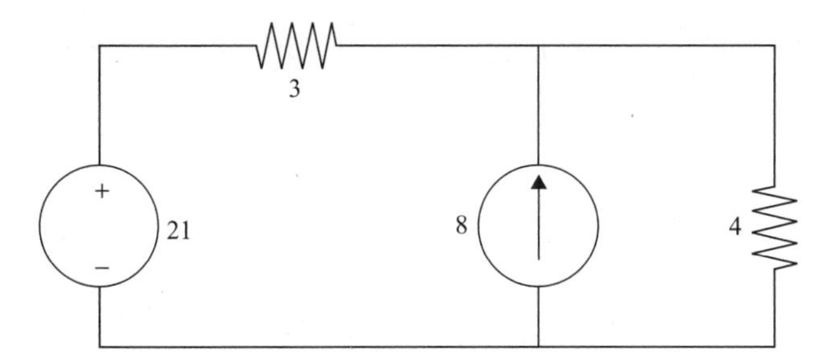

Fig. 4-4 When applying superposition to a circuit containing a current source, we will replace it by an open circuit.

If we combine them we get

$$P_1 + P_2 = R(I_1')^2 + R(I_1'')^2 = R[(I_1')^2 + (I_1'')^2]$$

The actual power in the resistor is

$$P = RI_1^2 = R(I_1' + I_1'')^2 = (I_1')^2 + 2I_1'I_1'' + (I_1'')^2$$

This shows that

$$P_1 + P_2 \neq P$$

So superposition can be used to find the correct current, but it cannot be used to find the power.

When using superposition in a circuit that contains current sources, we replace each current source by an open circuit. We will apply superposition to the circuit shown in Fig. 4-4 to see how this works.

EXAMPLE 4-2

Find the power absorbed by the 3 Ω resistor in Fig. 4-4. Use superposition to find the current flowing through the resistor.

SOLUTION

The first step is to set one of the sources to zero. We will set the current source to zero first and denote the current flowing through the 3 Ω resistor due to the voltage source alone by I_V. To set the current source to zero, we replace it by an open circuit. This is illustrated in Fig. 4-5.

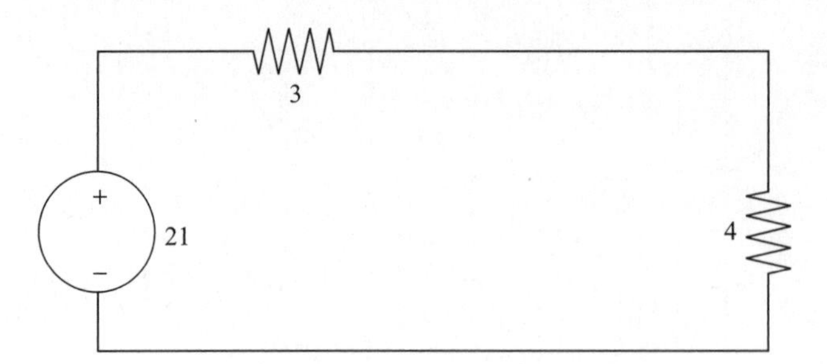

Fig. 4-5 Step one in solving the circuit in Fig. 4-4 using superposition is to replace the current source by an open circuit.

With two resistors in series, it is easy to find the current using a voltage divider. We find

$$I_V = \frac{21}{3+4} = 3 \text{ A}$$

where we have taken the current to be flowing in the clockwise direction. The next step is to set the voltage source to zero, replacing it by a short circuit while leaving the current source intact. This results in the circuit shown in Fig. 4-6.

This time we apply KCL at the top node where the wire from the current source meets the two resistors. We denote the current through the 3 Ω resistor due to the current source alone by I_C. This time a current divider gives

$$I_C = \frac{4}{4+3} = \frac{4}{7} \text{ A}$$

The total current flowing through the resistor is the sum

$$I_V + I_C = 3 + \frac{4}{7} = \frac{25}{7} \text{ A}$$

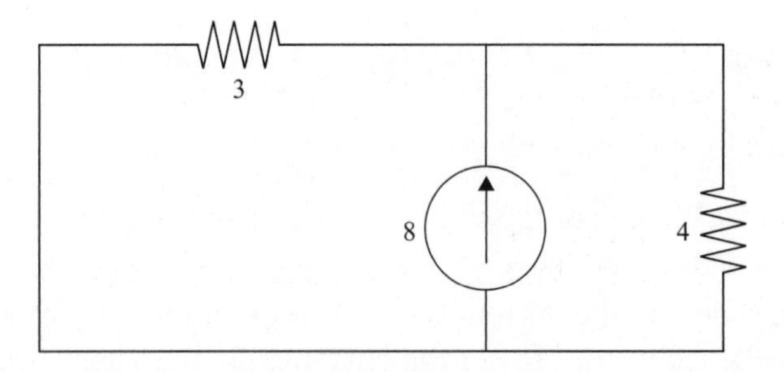

Fig. 4-6 The circuit in Fig. 4-4, with the voltage source set to zero.

Therefore, the power absorbed by the resistor is

$$P = RI^2 = (3)\left(\frac{25}{7}\right)^2 \approx 38 \text{ W}$$

Millman's Theorem

Consider a circuit containing a set of elements in parallel, where each element consists of a voltage source and resistor in series. We can replace such a circuit with a single voltage source and resistor in series by using *Millman's theorem.* This theorem is actually simple to apply. We just use the following steps:

1. Replace each voltage source V_i by an equivalent current source given by $I_i = G_i V_i$, where $G_i = 1/R_i$ is the conductance of the resistor in series with the voltage source V_i.
2. Compute the sum $\sum_i G_i V_i = GV = I$ to obtain a single current source. Sum up the conductances to obtain a single resistor in parallel with this current source.
3. Invert to obtain a single voltage source and resistor that are in series.

The *Millman voltage* is

$$V_M = \frac{G_1 V_1 + G_2 V_2 + \cdots + G_n V_n}{G_1 + G_2 + \cdots + G_n} \tag{4.2}$$

The *Millman resistance* is

$$R_M = \frac{1}{G_1 + G_2 + \cdots + G_n} \tag{4.3}$$

EXAMPLE 4-3

Consider the circuit shown in Fig. 4-7 and replace it by its Millman equivalent.

SOLUTION

First we replace the two voltage sources by equivalent current sources. In the first case consider the 5 V source. The conductance is found by inverting the resistance in series with the 5 V source, that is

$$G_1 = \frac{1}{3}$$

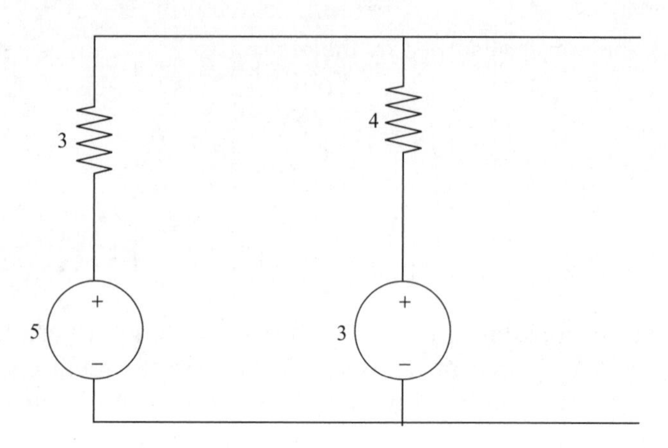

Fig. 4-7 In Example 4-3, we replace this circuit by a single voltage source and resistor in series.

The current source is found to be

$$I_1 = G_1 V_1 = \frac{1}{3}(5) = \frac{5}{3}$$

Next, we apply the same procedure to the second resistor–voltage source pair. The second conductance is

$$G_2 = \frac{1}{4}$$

The second current source is

$$I_2 = G_2 V_2 = \frac{1}{4}(3) = \frac{3}{4}$$

Now we have the circuit shown in Fig. 4-8, with current sources in parallel with the resistors.

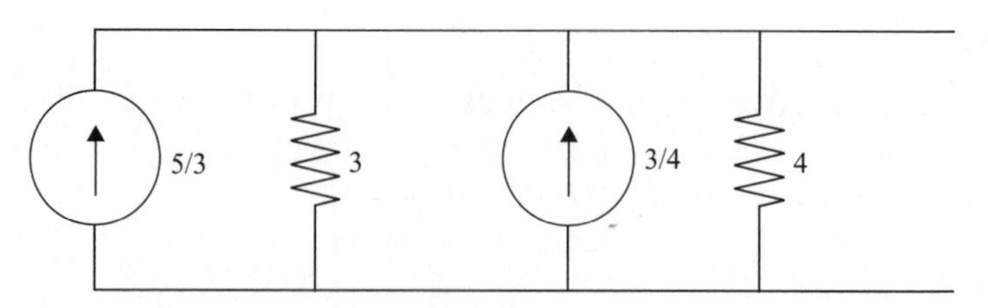

Fig. 4-8 The first intermediate step using Millman's theorem results in this circuit.

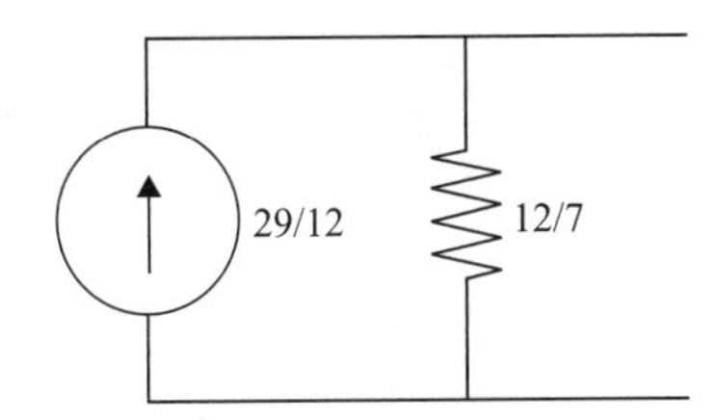

Fig. 4-9 Near the end of Millman's theorem, we have a single current source and resistor.

Now we add up the current sources to obtain a single current source given by

$$I = G_1 V_1 + G_2 V_2 = \frac{5}{3} + \frac{3}{4} = \frac{29}{12}$$

The total conductance is

$$G = G_1 + G_2 = \frac{1}{3} + \frac{1}{4} = \frac{7}{12}$$

We use these results to obtain the circuit shown in Fig. 4-9.

Now we can calculate the Millman voltage and Millman resistance. The Millman voltage is

$$V_M = \frac{G_1 V_1 + G_2 V_2}{G_1 + G_2} = \frac{29/12}{7/12} = \frac{29}{7} \text{ V}$$

$$R_M = \frac{1}{G_1 + G_2} = \frac{12}{7} \ \Omega$$

The Millman equivalent circuit, consisting of this voltage and resistance in series, is shown in Fig. 4-10.

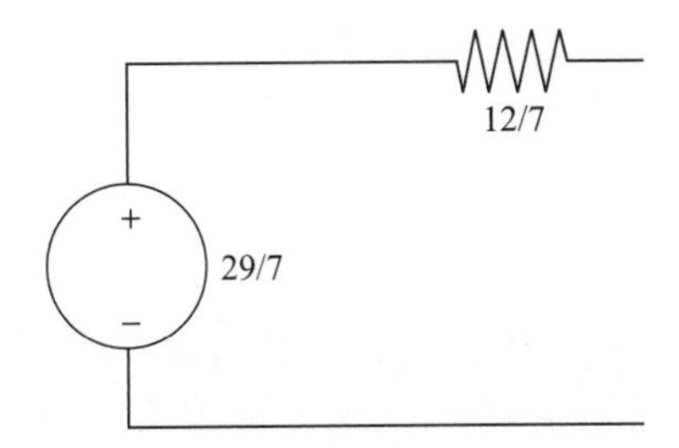

Fig. 4-10 The Millman equivalent circuit to the circuit shown in Fig. 4-7.

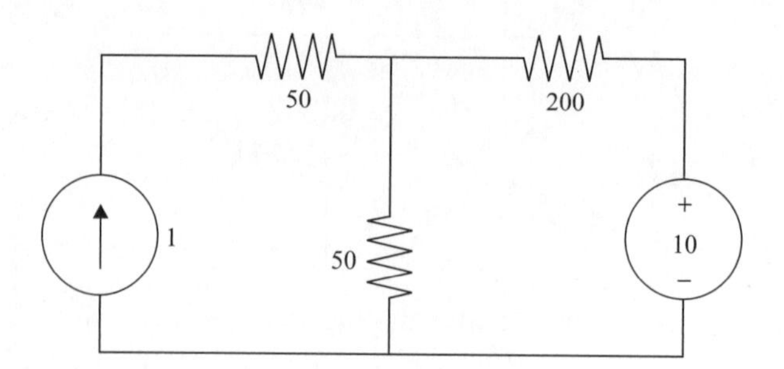

Fig. 4-11 Circuit for Problem 1.

Quiz

1. Find the voltage across the 1 A current source using superposition.
2. Can the power absorbed by the 50 Ω resistor in Fig. 4-11 be found by using superposition? If not, why not?
3. Write down the Millman equivalent voltage.
4. A circuit contains n voltage sources in parallel; each voltage source is in series with a resistor. What is the Millman equivalent resistance?
5. A circuit consists of components E_1, E_2, and E_3 in parallel. The first element E_1 is a 4 V voltage source and 2 Ω resistor in series. The second element E_2 is a 1 V voltage source and 1 Ω resistor in series, and the final element E_3 is a 2 V voltage source and 3 Ω resistor in series. Find the Millman voltage and the Millman resistance for this circuit.

Delta–Wye Transformations and Bridge Circuits

In this chapter we cover three common circuit configurations.

Delta–Wye Transformations

A *Y* or *wye* resistor circuit is a set of three resistors connected in a Y formation, as shown in Fig. 5-1.

A *delta* or Δcircuit is a set of three resistors connected in a Δ or triangular formation. This is shown in Fig. 5-2.

These two circuits can be transformed into each other, providing the notion of $\Delta - Y$*equivalence.* This equivalence works in a manner similar to Thevenin

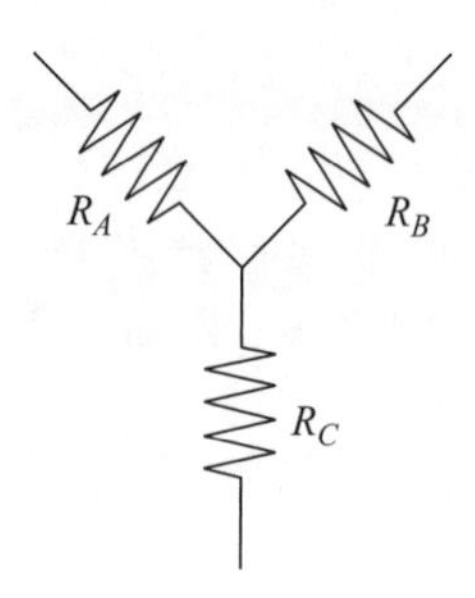

Fig. 5-1 Three resistors arranged in the Wye configuration.

equivalence, in that the two circuits look the same if seen by an *external* observer. However, they may be very different internally, so don't expect to find the same voltages or currents associated with any given pair of resistors. You can think of the equivalence as a black box that responds with the same voltage and current, but you don't know what components are inside or how they are effected individually.

First let's consider the transformation from a Y circuit (Fig. 5-1) to a Δ circuit (Fig. 5-2). Some tedious algebra can be used to derive the equivalent circuits, but we won't go through that and will just state the results. The first resistor in the Δconfiguration can be shown to be related to the Y circuit via

$$R_1 = \frac{R_A R_B + R_A R_C + R_B R_C}{R_B} \tag{5.1}$$

The second resistance in Fig. 5-2 can be calculated from the resistances in Fig. 5-1 to be

$$R_2 = \frac{R_A R_B + R_A R_C + R_B R_C}{R_C} \tag{5.2}$$

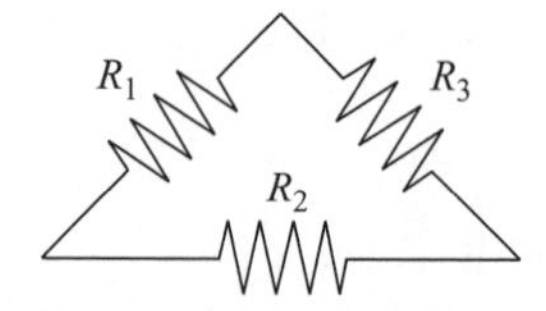

Fig. 5-2 Three resistors in the Δ configuration.

And finally, the third resistance is given by

$$R_3 = \frac{R_A R_B + R_A R_C + R_B R_C}{R_A} \tag{5.3}$$

Now let's consider the inverse or opposite transformation. Given the Δcircuit shown in Fig. 5-2, how can we derive the Y circuit shown in Fig. 5-1? There are three simple formulas that can once again be derived using tedious algebra. We will just list them. The first resistance is

$$R_A = \frac{R_1 R_2}{R_1 + R_2 + R_3} \tag{5.4}$$

Next we have

$$R_B = \frac{R_2 R_3}{R_1 + R_2 + R_3} \tag{5.5}$$

And finally, the third resistance in the Y configuration can be derived from the Δ configuration as

$$R_C = \frac{R_1 R_3}{R_1 + R_2 + R_3} \tag{5.6}$$

EXAMPLE 5-1
Convert the circuit shown in Fig. 5-3 into the Y configuration.
First we compute the sum of the individual resistances

$$R_T = R_1 + R_2 + R_3 = 2 + 4 + 7 = 13\ \Omega$$

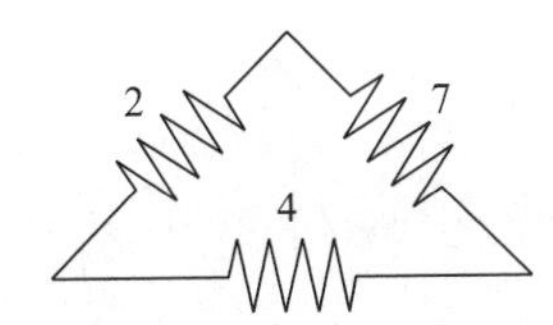

Fig. 5-3 The initial circuit for Example 5-1.

We denote the resistances in the Y configuration as they are shown in Fig. 5-1. Using (5.4) we find the first resistance to be

$$R_A = \frac{R_1 R_2}{R_T} = \frac{(2)(4)}{13} = \frac{8}{13}\ \Omega$$

The second resistance is found using (5.5)

$$R_B = \frac{R_2 R_3}{R_T} = \frac{(4)(7)}{13} = \frac{28}{13}\ \Omega$$

Finally, we use (5.6) to calculate the third resistance

$$R_C = \frac{R_1 R_3}{R_T} = \frac{(2)(7)}{13} = \frac{14}{13}\ \Omega$$

The transformed Y equivalent circuit to Fig. 5-3 is shown in Fig. 5-4.

EXAMPLE 5-2

Consider the circuits shown in Fig. 5-5. Can they be related by a Δ–Y transformation?

SOLUTION

First note that the resistor $R_A = 2$ in the Y configuration. Now using (5.4) and the Δcircuit shown in Fig. 5-5 consider that

$$R_A = \frac{R_1 R_2}{R_1 + R_2 + R_3} = \frac{(3)(4)}{3 + 4 + 2} = \frac{12}{9} \neq 2$$

Hence, these two circuits are not related by a Δ–Y transformation.

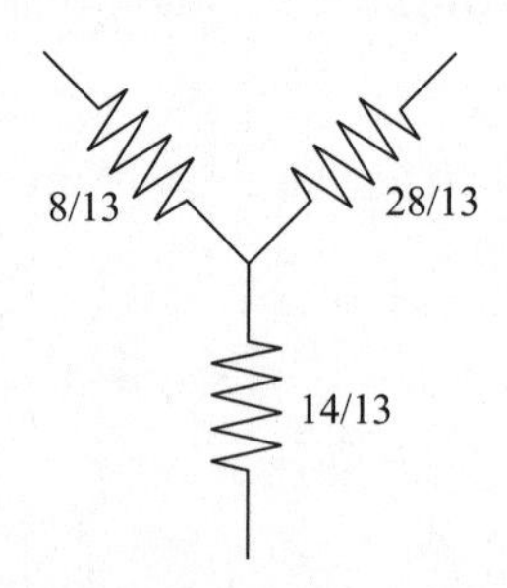

Fig. 5-4 This circuit was obtained from the circuit shown in Fig. 5-3 by using delta–wye equivalence.

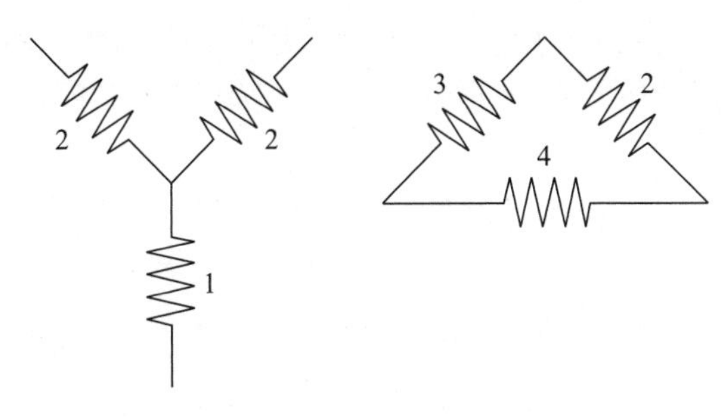

Fig. 5-5 A Y and Δ circuit for Example 5-2.

Hence, these two circuits are not related by a Δ–Y transformation.

Bridge Circuits

A *bridge* is a resistive circuit with two deltas or two Y's connected. This is shown in Fig. 5-6 where two Δ circuits share the same base resistor.

A *Wheatstone bridge* is a variation of this circuit which can be used to measure an unknown resistance. Looking at Fig. 5-6, we can construct a Wheatstone bridge in the following way. A voltage source V is connected to terminals A-B. Next, we replace R_5 with a Galvanometer (a device that can measure current) which is connected to the rest of the circuit by a switch which is initially open. The resistance R_2 is an adjustable resistance. An unknown resistance is placed in the position of R_4, the device is designed to determine what R_4 is.

The variable resistance R_2 is adjusted until the galvanometer switch can be closed without causing any movement in the galvanometer needle. This means that there is zero voltage across the galvanometer. At that time the voltages across the four resistors R_1, R_2, R_3, R_4 satisfy

$$V_1 = V_2, \quad V_3 = V_4$$

Fig. 5-6 A bridge circuit.

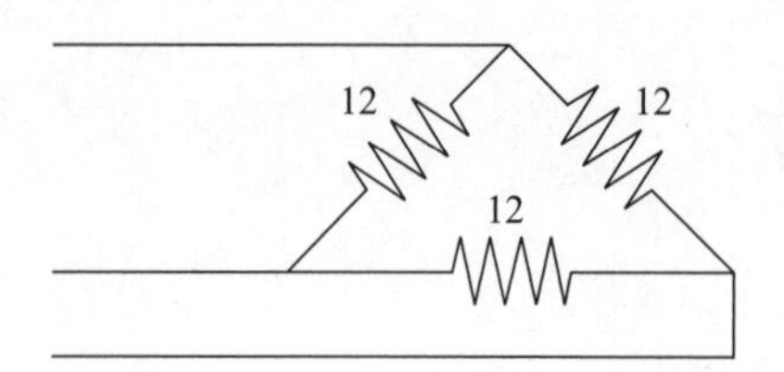

Fig. 5-7 A Δ configuration.

When this condition is met, the Wheatstone bridge is said to be *balanced*. Using voltage dividers we see that this condition translates into

$$\frac{R_1 V}{R_1 + R_3} = \frac{R_2 V}{R_2 + R_4} \tag{5.7}$$

And

$$\frac{R_3 V}{R_1 + R_3} = \frac{R_4 V}{R_2 + R_4} \tag{5.8}$$

Dividing (5.8) by (5.7) gives the bridge balance equation

$$R_4 = \frac{R_2 R_3}{R_1} \tag{5.9}$$

Quiz

1. Consider the circuit shown in Fig. 5-7 and convert it into an equivalent Y circuit.
2. Three resistors $R = 12$ are connected in a Y configuration. What is R for the equivalent Δ circuit?
3. In a Wheatstone bridge with $R_1 = 2$, $R_3 = 4$ it is found that balance is achieved when $R_2 = 6$. What is the value of the unknown resistance?

Capacitance and Inductance

So far we have looked at resistive circuit elements. In this chapter we extend our analysis to include two important electric devices: the capacitor and the inductor. The operation of these devices is more involved than what we have seen so far. In fact, as we will see shortly, the relationships between voltage and current involve calculus. This means that when we include these devices in our analysis, we mark the end of a purely algebraic approach and are faced with having to solve differential equations. We begin with the capacitor.

The Capacitor

A *capacitor* is a device that is capable of storing electric charge. It is not our purpose to discuss the specific physical nature or the construction of a capacitor. This information can be found in any basic physics book. Rather we will focus

Fig. 6-1 The representation of a capacitor as a circuit element.

on the behavior of capacitors in electric circuits. This means finding a voltage–current relation analogous to Ohm's law and determining how to calculate the power emitted or absorbed by a capacitor. We can then use this information to analyze electric networks that contain capacitors. The symbol used to denote a capacitor is shown below in Fig. 6-1.

The ability or *capacity* of a capacitor to store electric charge is measured in terms of charge per applied voltage. In SI units, capacitance is measured in Coulombs per volt, which are denoted by a special unit called the *Farad*. Specifically

$$C = \frac{Q}{V} \text{ [F]} \tag{6.1}$$

In most realistic situations, the capacitance is a very small value. Therefore, you will see capacitance on the order of microfarads or even picofarads. In some examples in this book, however, we use large values for instructional purposes.

Capacitors in Parallel or Series

Like resistance, we can form an equivalent capacitance when faced with a set of capacitors connected in parallel or in series. When a set of capacitors are connected in parallel, the total equivalent capacitance is found by adding up the individual capacitances. That is

$$C_T = C_1 + C_2 + C_3 + \cdots \tag{6.2}$$

If the capacitors are arranged in series, then

$$C_T = \frac{1}{1/C_1 + 1/C_2 + 1/C_3 + \cdots} \tag{6.3}$$

The alert reader will notice that capacitors and resistors connected in parallel and in series are added up in the opposite manner. Let's consider an example.

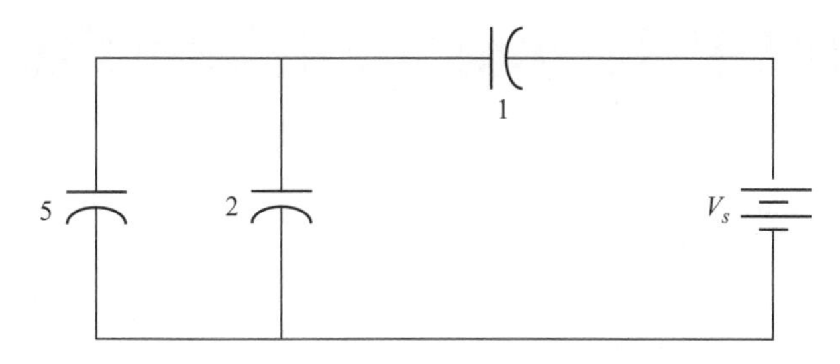

Fig. 6-2 Circuit analyzed in Example 5-1.

EXAMPLE 6-1

What is the total capacitance as seen by the voltage source in Fig. 6-2? All capacitances are given in microfarads.

SOLUTION

The 5 and 2 μF capacitors are in parallel. Hence they can be replaced by a single capacitor with

$$C' = 5 + 2 = 7\ \mu\text{F}$$

The circuit can be replaced by the circuit shown in Fig. 6-3.

The 7 and 1 μF capacitors are connected in series. Using (6.3) these two capacitors can be replaced by an equivalent capacitor with

$$C'' = \frac{1}{1/7 + 1} = 8/7\ \mu\text{F}$$

This is the total equivalent capacitance as seen by the voltage source.

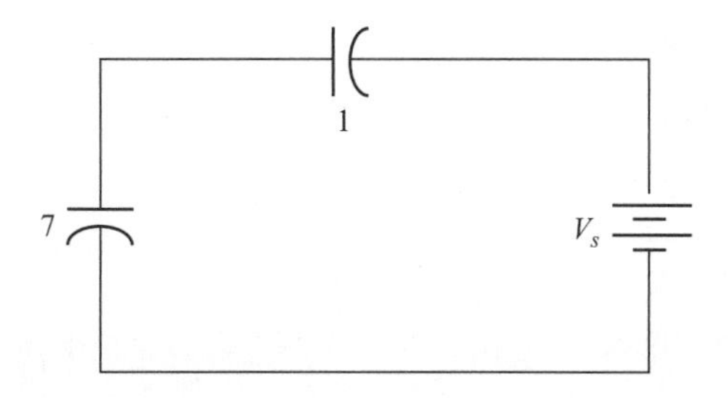

Fig. 6-3 Circuit obtained from that shown in Fig. 6-2 using the rule for capacitors connected in parallel.

Voltage–Current Relations in a Capacitor

The quantities that are of most interest in circuit analysis are the voltage and current in a circuit element. To see how these are related, we begin with the fundamental relation used to define capacitance, equation (6.1). Let's rearrange the terms a bit

$$CV = Q$$

Now we take the time derivative of this expression

$$C\frac{dV}{dt} = \frac{dQ}{dt}$$

Capacitance C is a constant. Now recall that current is the time rate of change of charge

$$i = \frac{dQ}{dt}$$

Therefore, we have found that the current flowing through a capacitor is related to the voltage across that capacitor in the following way

$$i(t) = C\frac{dv}{dt} \tag{6.4}$$

EXAMPLE 6-2
Find the voltage across each capacitor in the circuit shown in Fig. 6-4.

SOLUTION
We can attack this problem by first finding the equivalent capacitance seen by the voltage source. The 2 and the 3 μF capacitors are in parallel, so they

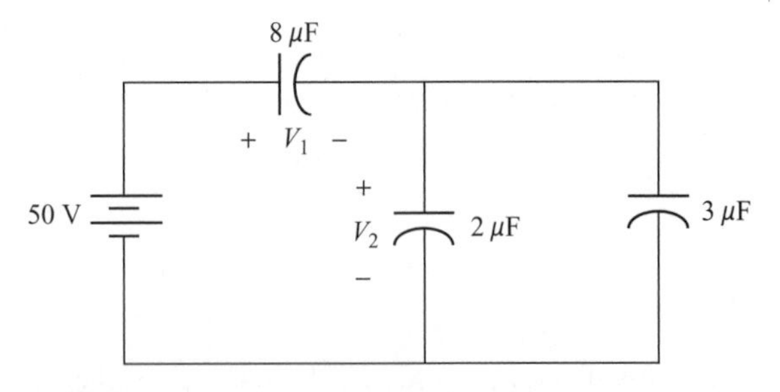

Fig. 6-4 Circuit analyzed in Example 6-2.

can be replaced by

$$C' = 2 + 3 = 5\ \mu\text{F}$$

Now we have the 8 μF capacitor in series with C'. Hence the total or equivalent capacitance seen by the voltage source is

$$C_T = \frac{1}{1/8 + 1/5} = \frac{40}{13} = 3.1\ \mu\text{F}$$

This capacitance will be in parallel with the voltage source, so the voltage across the equivalent capacitance is 50 V. This allows us to determine the total charge in Coulombs as

$$Q = CV = (3.1\ \mu\text{F})(50\ \text{V}) = 155\ \mu\text{C}$$

This is the charge on the 8 μF capacitor in the original circuit. So the voltage V_1 in Fig. 6-4 is

$$V_1 = \frac{155 \times 10^{-6}\ \text{C}}{8 \times 10^{-6}\ \text{F}} = 19.4\ \text{V}$$

Now we can use KVL to find the voltage across the 2 and 3 μF capacitors. We have

$$-50 + 19.4 + V_2 = 0$$
$$\Rightarrow V_2 = 30.6\ \text{V}$$

Voltage in Terms of Current

Equation (6.4) tells us how to find the current in terms of the applied voltage for a capacitor. This relation can be inverted to give the voltage in terms of the current as a function of time by integrating. Specifically, if the initial voltage is denoted by $v(0)$ we have

$$v(t) = v(0) + \frac{1}{C}\int_0^t i(s)\,ds \tag{6.5}$$

In (6.5), s is just a dummy variable of integration. Let's see how this equation works with some examples.

EXAMPLE 6-3

It is known that when $t \geq 0$ the current flowing through a 5 F capacitor is described by $i(t) = 3e^{-2t}$ [A]. If the initial voltage is zero, find and plot the voltage across the capacitor as a function of time. What is the charge on the capacitor as a function of time?

SOLUTION

Using (6.5) with $v(0) = 0$ we have

$$v(t) = \frac{1}{5}\int_0^t 3e^{-2s}\,ds = -\frac{3}{10}e^{-2s}\Big|_0^t = -\frac{3}{10}e^{-2t} + \frac{3}{10} = \frac{3}{10}(1 - e^{-2t}) \text{ [V]}$$

A plot of the voltage is shown in Fig. 6-5. What should you notice about the plot? The important characteristic of the plot is that the voltage rises up to a constant value, when it attains $v(t) = 0.3$ V. When the voltage becomes constant, notice, by looking at (6.4), that no more current flows through the capacitor.

To find the charge on the capacitor as a function of time, we apply equation (6.1)

$$q(t) = Cv(t) = 5\left\{\frac{3}{10}(1 - e^{-2t})\right\} = \frac{3}{2}(1 - e^{-2t})\text{ C}$$

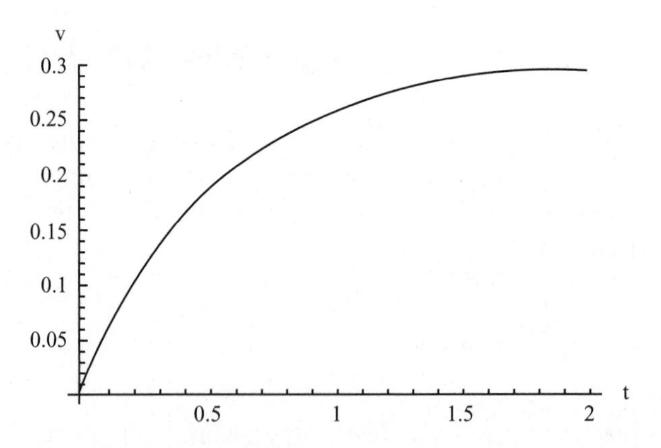

Fig. 6-5 The voltage in Example 5-3 reaches a constant value in about 2 s.

Power and Energy in the Capacitor

The instantaneous power in any circuit element can be found using first principles. Recall that

$$p(t) = v(t)i(t)$$

To find the power in a capacitor, we simply apply (6.4). This tells us that

$$p(t) = Cv(t)\frac{dv}{dt} \tag{6.6}$$

The energy is found by integrating

$$w = \int_{t_1}^{t_2} p(t)\,dt = C\int_{v_1}^{v_2} v\,dv \tag{6.7}$$

where v_1 is the voltage at time t_1, and do on. So

$$w = \frac{1}{2}Cv^2 \tag{6.8}$$

EXAMPLE 6-4

Suppose that the current flowing through a 2 F capacitor is $i(t) = 2e^{-3t}$. Find the energy in the capacitor for $0 \le t \le 2$ s. The initial voltage across the capacitor is 1 V.

SOLUTION

First we find the voltage as a function of time

$$v(t) = 1 + \frac{1}{2}\int_0^t 2e^{-3s}\,ds = 1 - \frac{1}{3}e^{-3t} + \frac{1}{3} = \frac{4}{3} - \frac{1}{3}e^{-3t} \text{ V}$$

The instantaneous power is

$$p(t) = v(t)i(t) = \left(\frac{4}{3} - \frac{1}{3}e^{-3t}\right)2e^{-3t} = \frac{8}{3}e^{-3t} - \frac{2}{3}e^{-6t} \text{ W}$$

The energy for $0 \le t \le 2$ is found by integration

$$w = \int_0^2 p(t)\,dt = \int_0^2 \frac{8}{3}e^{-3t} - \frac{2}{3}e^{-6t}\,dt = \frac{1}{9}(7 + e^{-12} - 8e^{-6}) = 0.78\text{ J}$$

Time Constants, Zero-Input Response, and First-Order RC Circuits

We call a circuit consisting of a resistor and capacitor an *RC* circuit. Initially, the capacitor is connected to a voltage source by a switch. This allows the capacitor to charge up and attain an initial voltage. When the switch is thrown, the connection between the capacitor and the voltage source is broken and the capacitor is connected to the resistor. When this happens, the elements in the RC circuit start off with initial voltages and currents that decay rapidly to zero as the capacitor discharges through the resistor. In fact, the voltages and currents decay *exponentially*. We call such rapidly decaying voltages and currents *transients*.

The initial circuit is shown in Fig. 6-6, where we see the capacitor connected to a voltage source. There is an open circuit between the capacitor and the resistor so nothing is happening to the resistor.

Initially, it is trivial to see that applying KVL to the closed loop containing the voltage source and the capacitor puts a voltage V across the capacitor. This will be the initial condition when the switch is thrown to the right, disconnecting the voltage source and connecting the capacitor to the resistor. When this is done, we are left with the circuit shown in Fig. 6-7.

To solve this circuit, we can apply KCL. Take a node at the top of the resistor and suppose that the current is flowing in the clockwise direction. If we take +

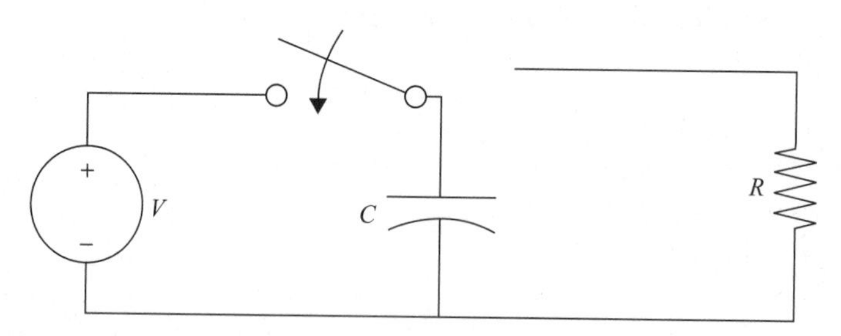

Fig. 6-6 First we charge up the capacitor by connecting it via a switch to a voltage source V.

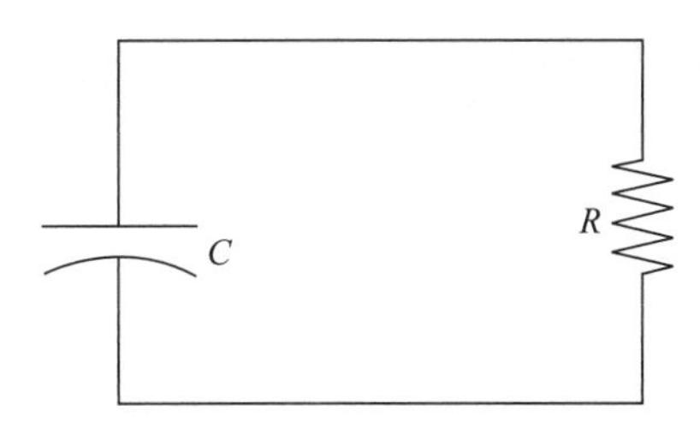

Fig. 6-7 The switch in Fig. 6-6 is thrown to the right, giving us a simple RC circuit.

for currents entering the node and use (6.4) for the current flowing through the capacitor, we have

$$i(t) = C\frac{dv}{dt} = -i_R(t) \tag{6.9}$$

The current flowing through the resistor can be written in terms of the voltage as $i_R(t) = v_R(t)/R$. Looking at Fig. 6-7 one can see immediately that the same voltage must be across the capacitor and the resistor (KVL), so (6.9) can be written as

$$C\frac{dv}{dt} = -\frac{v}{R} \tag{6.10}$$

This is an easy differential equation to solve, but we will go through it in a bit of detail for readers who are new to the material or who are simply rusty. First we move all terms involving v to the left side of the equation and all other terms to the right side of the equation. This gives

$$\frac{dv}{v} = -\frac{1}{RC}dt$$

If you look in an integral table you will find that

$$\int \frac{dx}{x} = \ln x + C$$

Hence, ignoring the constant of integration and initial conditions (which will be put on the right side), if we integrate the left-hand side we obtain

$$\int \frac{dv}{v} = \ln v(t)$$

Now let's integrate the right-hand side. This is elementary

$$-\frac{1}{RC}\int dt = -\frac{t}{RC} + K$$

So we have the following relationship

$$\ln v(t) = -\frac{t}{RC} + K$$

Now we use the fact that $e^{\ln x} = x$ and exponentiate both sides giving

$$v(t) = K\,e^{-t/RC}$$

Notice that

$$v(0) = K$$

Hence, the constant of integration is the initial voltage, which we had labeled V in Fig. 6-6. Let's call it $v(0)$, however, to emphasize that this is the initial condition for the time-dependent voltage function. So we have found that the voltage in an RC circuit is given by

$$v(t) = v(0)e^{-t/RC} \tag{6.11}$$

Looking at this decaying exponential it's easy to see that the voltage quickly dissipates, dying away to zero in short time. This reflects the fact that this voltage is a transient. Later, when we consider sinusoidal sources, we will focus on the long-term solutions that are called *steady-state* solutions.

Returning to our example, notice that when $t = RC$ the voltage has decayed to a fraction $1/e = e^{-1}$ of its initial value. The time $\tau = RC$ is called the *time constant* of the circuit. When $t = \tau$, the voltages and currents in the circuit are just 36.8% of their initial value. At $t = 5\tau$, or at five time constants, the voltages and currents are a mere 0.6% of their initial values. So a good rule of thumb to remember is that at five time constants the voltages and currents in an RC circuit are essentially zero.

EXAMPLE 6-5

A capacitor $C = 1/10$ Fis in an RC circuit with a resistor $R = 5\ \Omega$ as shown in Fig. 6-7. What is the time constant for this circuit?

SOLUTION

The time constant is found by multiplying the resistance and capacitance together. In this case

$$\tau = (5\ \Omega)(1/10\ \text{F}) = 0.5\ \text{s}$$

This tells us that the voltage across each element will be 36.8% of the initial value in half a second. In 2.5 s, the voltage will be almost zero.

EXAMPLE 6-6

The voltage in an RC circuit is given by (6.11). Suppose that $v(0) = 10$ V, $R = 50\ \Omega$, and $C = 0.3$ F. How long will it take for the voltage across the capacitor to decay to 3 V? What current initially flows through the circuit?

SOLUTION

The solution for the voltage in an RC circuit is given by (6.11). In this case,

$$v(t) = 10e^{-t/15}$$

The time constant is

$$\tau = RC = (50\ \Omega)(0.3\ \text{F}) = 15\ \text{s}$$

To determine how long it will take for the voltage to decay to 3 V, we need to solve the equation

$$3 = 10e^{-t/15}$$

Rearranging

$$\frac{3}{10} = e^{-t/15}$$

Taking the natural logarithm of both sides we find

$$\ln\left(\frac{3}{10}\right) = -t/15,$$
$$\Rightarrow t = -15\ln\left(\frac{3}{10}\right) = -18.06\ \text{s}$$

So it takes about 18 s for the voltage to decay from the initial 10 V down to 3 V. Note that since the time constant is 15 s, it will take about $5\tau = 75$ s or 1 min 15 s for the voltages and currents in this circuit to dissipate to zero.

The current is found by using (6.4)

$$i(t) = C\frac{dv}{dt} = (0.3)\frac{d}{dt}(10e^{-t/15}) = 3\left(-\frac{1}{15}\right)e^{-t/15} = -\frac{1}{5}e^{-t/15}$$

Hence, at $t = 0$, a current of $1/5 = 0.2$ A flows in a counterclockwise direction in the circuit shown in Fig. 6-7.

When we do circuit analysis, we can draw a circuit without voltage and current sources as in Fig. 6-7 and just allow the voltages and currents to have *initial conditions*. When we analyze a circuit in this way, we call this the *zero-input response* of the circuit.

The Inductor

Now we meet our second dynamic circuit element, the *inductor*. An inductor is a wire wound into a coil. The symbol used to represent an inductor in a circuit is shown in Fig. 6-8.

From elementary physics, we know that a current flowing through a straight wire produces a magnetic field, with circular lines of magnetic force about the wire. We measure the density of these field lines and call them *magnetic flux*. Winding a wire into a coil increases the magnetic flux.

For our purposes, an inductor can be thought of as a dynamic circuit element that stores electric energy. The key equation when working with inductors relates the voltage across it to the current flowing through it via

$$v = L\frac{di}{dt} \tag{6.12}$$

The constant L is a property of the inductor itself called the *inductance*. We measure inductance in *henries* (H).

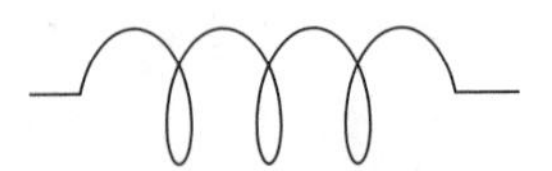

Fig. 6-8 An inductor.

Inductors in Series and in Parallel

If n inductors are connected in series, the total inductance is given by

$$L = L_1 + L_2 + \cdots + L_n \tag{6.13}$$

If a set of n inductors is connected in parallel, the total inductance is

$$L_T = \frac{1}{\dfrac{1}{L_1} + \dfrac{1}{L_2} + \cdots + \dfrac{1}{L_n}} \tag{6.14}$$

Energy in an Inductor

The energy in an inductor can be found by integrating the power. Going back to basics, the power $p = vi$. Using (6.12), we have

$$w = \int vi\,dt = \int L\frac{di}{dt}i\,dt = L\int i\,di$$

Integrating we find that the energy stored in an inductor is

$$w = \frac{1}{2}Li^2 \tag{6.15}$$

The energy is given in joules. At this point, it's a good idea to go back and compare the basic voltage–current relation (6.12) and energy (6.15) for the inductor with the equations we found for the capacitor, (6.4) and (6.8). Notice that these equations have similar form, we just interchange L and C and v and i.

Current in an Inductor

If we know the voltage in an inductor, we can find the current by integrating. The result is

$$i(t) = i(0) + \frac{1}{L}\int_0^t v(\tau)\,d\tau \tag{6.16}$$

Zero-Input Analysis of First-Order RL Circuits

First-order RL circuits can be analyzed in a similar manner that we used to analyze first-order RC circuits in the previous sections. We won't worry about sources but will provide initial conditions; hence, we will perform zero-input analysis.

EXAMPLE 6-7

A current $i(t) = 2e^{-t}$ flows through a 2 H inductor. What is the voltage across the inductor? Plot the voltage and current on the same graph. Calculate and plot the energy stored in the inductor as a function of time.

SOLUTION

We find the voltage across the inductor by using (6.12). This gives

$$v(t) = L\frac{di}{dt} = (2)\frac{d}{dt}2e^{-t} = -4e^{-t}$$

The voltage and current are shown together in Fig. 6-9.

The energy stored in the inductor is found readily from (6.15)

$$w = \frac{1}{2}Li^2 = \frac{1}{2}(2)(2e^{-t})^2 = 4e^{-2t}\text{ J}$$

A plot of the energy is shown in Fig. 6-10.

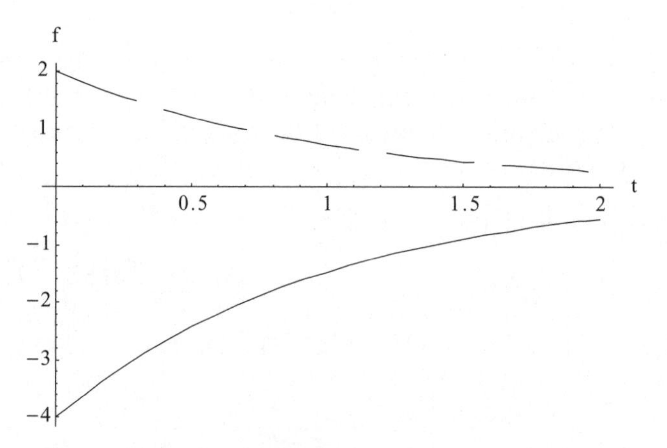

Fig. 6-9 A plot of the current (dashed line) and voltage (solid line) for the inductor in Example 6-7.

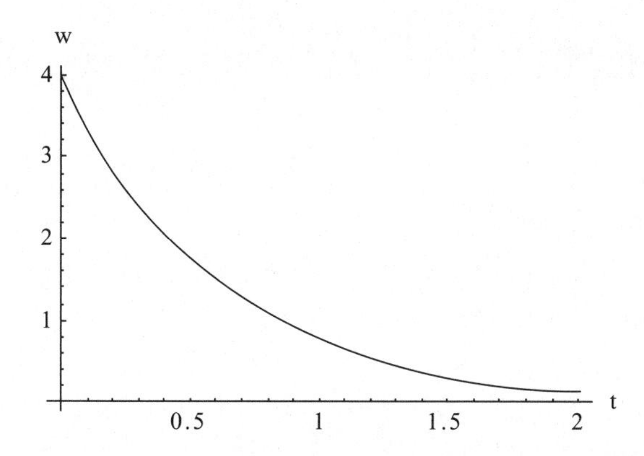

Fig. 6-10 The energy stored in the inductor in Example 6-7. Initially, 4 J are stored in the inductor. This dissipates in about 2 s.

Mutual Inductance

When two inductors are in proximity, the fact that the magnetic field lines extend out into space around the wires means that some flux from each wire will "link" the other wire. That is, a current flowing in one inductor will generate magnetic flux that links the other inductor and *induce* a current to flow through it. One way to bring two inductors into proximity is to wind two wires around the same iron core.

Let two inductors be identified by inductances L_1 and L_2 and be brought into proximity as shown in Fig. 6-11.

Each inductor has two types of flux, *leakage flux* and *mutual flux*. For inductor 1, we denote these by ϕ_{l1} and ϕ_{m1}, respectively, and similarly for the second inductor. We can measure how tightly the inductors are coupled by using the

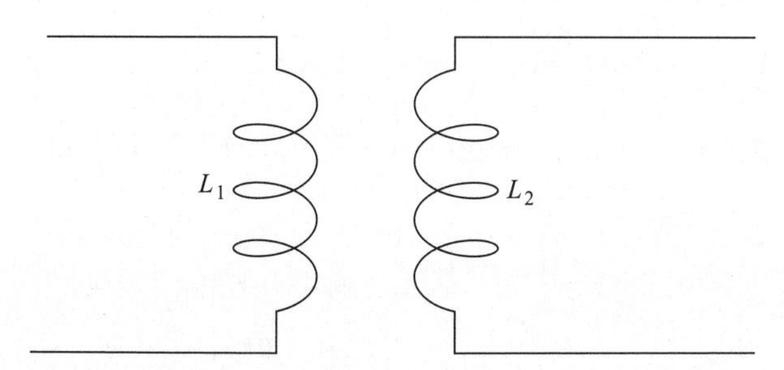

Fig. 6-11 Two inductors brought close to each other will have mutual inductance.

coefficient of coupling, which is defined by

$$k = \sqrt{\left(\frac{\phi_{m1}}{\phi_{l1} + \phi_{m1}}\right)\left(\frac{\phi_{m2}}{\phi_{l2} + \phi_{m2}}\right)} \tag{6.17}$$

The more tightly coupled they are, the higher k with $0 \le k \le 1$. The *self-inductance* L for a winding with N turns is given by

$$L = \frac{N(\phi_m + \phi_l)}{i(t)} \tag{6.18}$$

Now, magnetic flux is related to the current in an inductor via

$$\phi(t) = Li(t) \tag{6.19}$$

where ϕ is measured in *webers* (Wb). When the two inductors are brought into proximity, some flux will be induced in each inductor due to the flux of the other inductor. That is, some flux will be induced due to the current flowing in the other inductor. The induced flux is given by

$$\phi_i(t) = Mi_j(t)$$

Here we are saying that this is the induced flux in inductor i due to the current flowing in inductor j. We call the constant of proportionality M the *mutual inductance.* The number of windings in one inductor multiplied by the mutual flux in the second inductor, divided by the current in the second inductor, define the mutual inductance. That is,

$$M = \frac{N_1\phi_{m2}}{i_2(t)} = \frac{N_2\phi_{m1}}{i_1(t)} \tag{6.20}$$

When mutual inductance is present, we have to modify the equation relating voltage and current (6.12). Let inductor 1 have inductance L_1, voltage $v_1(t)$ with a current $i_1(t)$ flowing through it, and similarly for inductor 2. Then

$$v_1(t) = L_1\frac{di_1}{dt} + M\frac{di_2}{dt} \tag{6.21}$$

$$v_2(t) = L_2\frac{di_2}{dt} + M\frac{di_1}{dt} \tag{6.22}$$

In terms of mutual inductance, we can write the coefficient of coupling (6.17) as

$$k = \frac{M}{\sqrt{L_1 L_2}} \tag{6.23}$$

EXAMPLE 6-8

Two inductors are brought into proximity as shown in Fig. 6-11, with $L_1 = 0.5$ H and $L_2 = 0.7$ H. If a current flowing through inductor 2 is $i_2(t) = 20 \sin 100t$, and the induced voltage across inductor 1 is $v_1(t) = 700 \cos 100t$, what is the mutual inductance? Assume that initially no current is flowing through inductor 1.

SOLUTION

Using (6.21) with $i_1(t) = 0$ we have

$$v_1(t) = M\frac{di_2}{dt}, \Rightarrow$$

$$M = \frac{700 \cos 100t}{2000 \cos 100t} = 0.35$$

Since

$$\frac{di_2}{dt} = \frac{d}{dt}20 \sin 100t = 2000 \cos 100t$$

EXAMPLE 6-9

For the two inductors shown in Fig. 6-11, let $L_1 = 2$ H, $L_2 = 5$ H, and $M = 0.25$. Suppose that a current $i_2(t) = 100e^{-5t}$ flows through the second inductor. What is the voltage induced across L_1 if $i_1(t) = 0$? What is the coefficient of coupling between these two inductors?

SOLUTION

Using (6.21) we find

$$v_1(t) = M\frac{di_2}{dt} = (0.25)\frac{d}{dt}100e^{-5t} = -125e^{-5t} \text{ V}$$

The coefficient of coupling can be found from (6.23). We have

$$k = \frac{M}{\sqrt{L_1 L_2}} = \frac{0.25}{\sqrt{(2)(5)}} = 0.08$$

The inductors are not strongly coupled.

Zero-Input Response in an RL Circuit

Finding the zero-input response of an RL circuit proceeds in an fashion analogous to method used to find the zero-input response for an RC circuit. However, this time we take the initial condition to be the initial current rather than the initial voltage. The circuit we wish to solve is shown in Fig. 6-12.

Using

$$v = L\frac{di}{dt}$$

for the inductor and

$$v = Ri$$

For the resistor, we apply KVL to a loop about the circuit shown in Fig. 6-12. This gives the differential equation

$$L\frac{di}{dt} + Ri = 0$$

Hence

$$\frac{di}{i} = -\frac{R}{L}dt$$

Integrating gives us the transient current in the RL circuit. We find it to be

$$i(t) = i(0)e^{-(R/L)t} \tag{6.24}$$

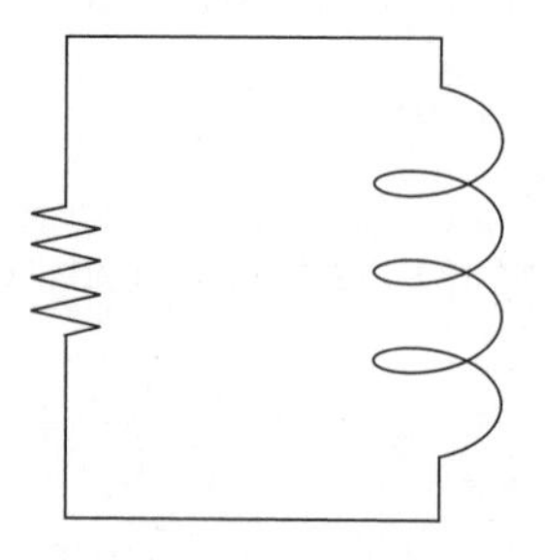

Fig. 6-12 Zero-input response of an RL circuit.

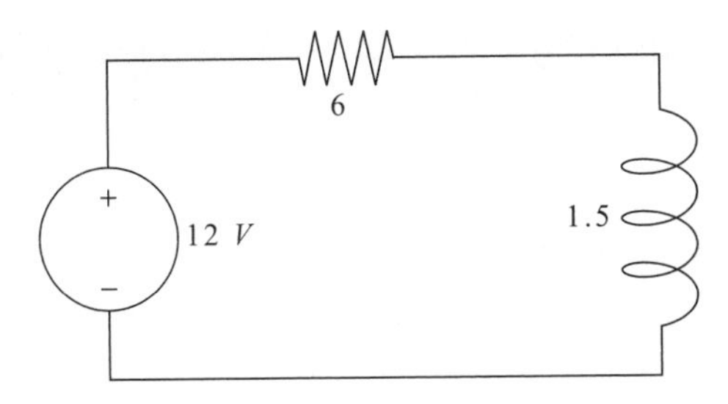

Fig. 6-13 A series RL circuit with a voltage source.

where $i(0)$ is the initial current. For an RL circuit, the time constant is given by

$$\tau = L/R \tag{6.25}$$

EXAMPLE 6-10

For the circuit shown in Fig. 6-13, find the total response of the current and the time constant. Plot the current as a function of time. Suppose that the initial current is 1 A.

SOLUTION

We apply KVL in a clockwise fashion starting at the voltage source. We have

$$-12 + 6i + 1.5\frac{di}{dt} = 0$$

Cleaning up a bit gives

$$\frac{di}{dt} + 4i = 8$$

This is an inhomogeneous differential equation—the presence of the extra term is due to the voltage source. When we solve for the total solution, we have a solution that includes the zero-input response + zero-state response. We have already seen that the zero-input response is the solution of a circuit when initial conditions are supplied but no sources are included in the circuit. The zero-state response is the response of the circuit due to the presence of sources. Mathematically, the zero-input response is the homogeneous solution to the differential equation describing the circuit, and the zero-state response is the inhomogeneous solution. In our case, we find the zero-input response by solving

$$\frac{di}{dt} + 4i = 0$$

We already have quite a bit of practice solving this equation. The solution is

$$i_H(t) = Ke^{-4t}$$

where H indicates this is the homogeneous part of the solution. We found 4 from the time constant, which is

$$\tau = L/R = \frac{1.5}{6} = 0.25 \text{ s}$$

We leave the initial condition undetermined, because we need the total solution to find it. The solution for the zero-state response is found from the particular solution of the differential equation. In our case this is relatively easy, since the voltage source is dc. So we try

$$i_p(t) = A$$

where A is a constant. The derivative is clearly zero, so we find A by plugging into

$$\frac{di}{dt} + 4i = 8$$

And solving

$$4A = 8$$

Hence

$$i_p(t) = 2$$

The total solution is

$$i(t) = i_p(t) + i_H(t) = 3 + Ke^{-4t}$$

With $i(0) = 1$ A, we have

$$1 = 2 + K, \Rightarrow K = -1$$

So the total solution with initial condition is

$$i(t) = i_p(t) + i_H(t) = 3 - e^{-4t}$$

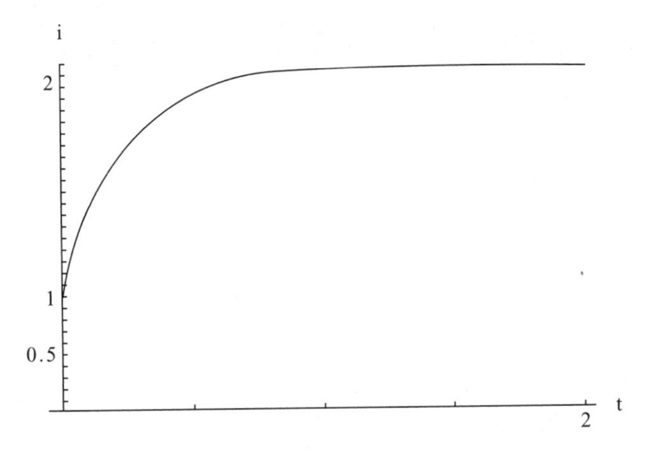

Fig. 6-14 A plot of the current found in Example 6-10.

The plot, shown in Fig. 6-14, shows the current approaching the final value of 2 A in about 1.25 s, as would be expected from the time constant.

EXAMPLE 6-11

A resistor $R = 12\ \Omega$ and inductor $L = 4$ H are connected in series with a sinusoidal voltage source $v_s(t) = 20 \sin 30t$. Find the total response of the circuit for $i(t)$. The initial current is zero.

SOLUTION

The differential equation describing the circuit is

$$4\frac{di}{dt} + 12i(t) = 20 \sin 30t$$

The zero-input response is found by setting the value due to the source (the input on the right-hand side of the equation) to zero. This gives

$$\frac{di_H}{dt} + 3i_H(t) = 0$$

We find the solution to be

$$i_H(t) = Ke^{-3t}$$

where K is a constant to be determined from the initial condition. Now we need to solve for the zero-state response, or the particular solution for the current.

The inhomogeneous term is

$$20 \sin 30t$$

Suggesting a solution of the form

$$i_p(t) = A \sin(30t + \phi)$$

The derivative of this expression is

$$\frac{di_p}{dt} = 30A \cos(30t + \phi)$$

Now use the following trig identities

$$\sin(x + y) = \sin x \cos y + \cos x \sin y$$
$$\cos(x + y) = \cos x \cos y - \sin x \sin y$$

To write

$$i_p(t) = A \sin(30t + \phi) = A \sin 30t \cos \phi + A \cos 30t \sin \phi$$
$$\frac{di_p}{dt} = 30A \cos(30t + \phi) = 30A \cos 30t \cos \phi - 30A \sin 30t \sin \phi$$

Then we have

$$\frac{di_p}{dt} + 3i_p(t) = 30A \cos 30t \cos \phi - 30A \sin 30t \sin \phi + 3A \sin 30t \cos \phi + 3A \cos 30t \sin \phi$$

Grouping terms and setting equal to the inhomogeneous term due to the voltage source we have

$$A \cos 30t(30 \cos \phi + 3 \sin \phi) + A \sin 30t(3 \cos \phi - 30 \sin \phi) = 5 \sin 30t$$

Since there is no term involving $\cos 30t$ on the right-hand side, it must be true that

$$30 \cos \phi + 3 \sin \phi = 0$$

This leads to the relation

$$\tan\phi = -10$$
$$\Rightarrow \phi = -84^\circ$$

Now, $3\cos(-84^\circ) - 30\sin(-84^\circ) = 3(0.105) - 30(-0.995) = 30.155$. So we can solve for A. Given that $30\cos\phi + 3\sin\phi = 0$, we are left with

$$A\sin 30t(3\cos\phi - 30\sin\phi) = 5\sin 30t$$
$$\Rightarrow$$
$$A = \frac{5}{3\cos\phi - 30\sin\phi} = 0.166$$

So the particular solution, which represents the zero-state solution, is

$$i_p(t) = 0.166\sin(30t - 84^\circ)$$

The total solution is

$$i(t) = i_p(t) + i_H(t) = 0.166\sin(30t - 84^\circ) + Ke^{-3t}$$

Setting this equal to zero gives

$$K = -0.166\sin(-84^\circ) = +0.166$$

The total solution is therefore

$$i(t) = 0.166(\sin(30t - 84^\circ) + e^{-3t})$$

Second-Order Circuits

A *second-order circuit* is one that includes capacitors and inductors in a single circuit. Such a circuit is called second order because of the nature of the current-voltage relations for inductors and capacitors. When the analysis is done, a second-order differential equation for the current or the voltage will result.

Second-order circuits are subject to a phenomenon known as *damping*. Before looking specifically at electric circuits, consider an arbitrary second-order differential equation of the form:

$$s^2 + 2\zeta\omega_n s + \omega_n^2 = 0$$

where ω_n is the *undamped natural frequency* and ζ is the *damping ratio*. The dynamic behavior of the system is then described in terms of these two parameters ζ and ω_n. To determine the behavior of a system, we look at the damping ratio. There are three possibilities:

If $0 < \zeta < 1$, the roots are complex conjugates and the system is underdamped and oscillatory.
If $\zeta = 1$, the roots of the system are equal and the response is critically damped.
If $\zeta > 1$, the roots are negative, real, and unequal. The system response is overdamped.

With this in mind, consider the RLC circuit second-order equation (6.28) that is specific for a resistor, inductor, and capacitor in series:

$$s^2 + \frac{R}{L}s + \frac{1}{LC} = 0$$

We see that $\omega_n^2 = \dfrac{1}{LC}$

And $2\zeta\omega_n = \dfrac{R}{L} \Rightarrow \zeta = \dfrac{R\sqrt{LC}}{2L}$

This provides us with the damped case relationships for an RLC second-order equation:

If $0 < \frac{R\sqrt{LC}}{2L} < 1$, then the roots are complex conjugates and the system is underdamped and oscillatory.
If $\frac{R\sqrt{LC}}{2L} = 1$, then the roots of the system are equal and the response is critically damped.
If $\frac{R\sqrt{LC}}{2L} > 1$, then the roots are negative, real, and unequal. The system response is overdamped.

As an example, we consider a series RLC circuit shown in Fig. 6-15.
If we apply KVL around the circuit, we obtain

$$Ri + L\frac{di}{dt} + v_c = V$$

The same current flows through each circuit element. In the capacitor, the relation

$$i(t) = C\frac{dv_c}{dt}$$

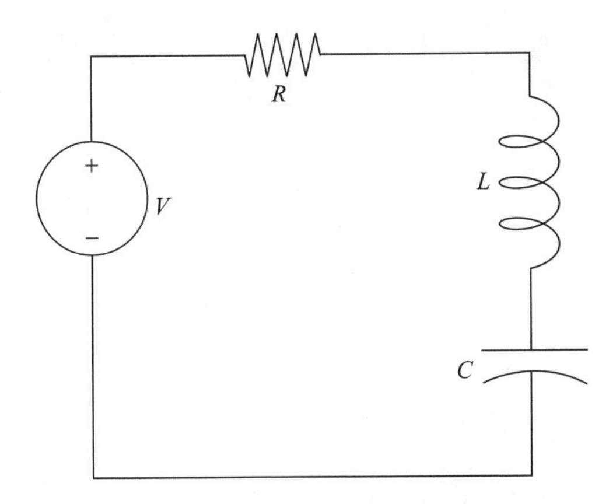

Fig. 6-15 A series RLC circuit.

is satisfied. Therefore, since the same current flows through the inductor, the voltage across the inductor is

$$L\frac{di}{dt} = LC\frac{d^2 v_c}{dt^2}$$

Also, the same current flows through the resistor, meaning that the voltage across the resistor is

$$Ri = RC\frac{dv_c}{dt}$$

Hence, in a series an RLC circuit with voltage source V is described by the differential equation

$$LC\frac{d^2 v_c}{dt^2} + RC\frac{dv_c}{dt} + v_c = V \tag{6.26}$$

The homogeneous solution or the zero-input solution is found by setting the source equal to zero

$$\frac{d^2 v_c}{dt^2} + \frac{R}{L}\frac{dv_c}{dt} + \frac{1}{LC}v_c = 0 \tag{6.27}$$

To solve this system, we write down the characteristic equation

$$s^2 + \frac{R}{L}s + \frac{1}{LC} = 0 \tag{6.28}$$

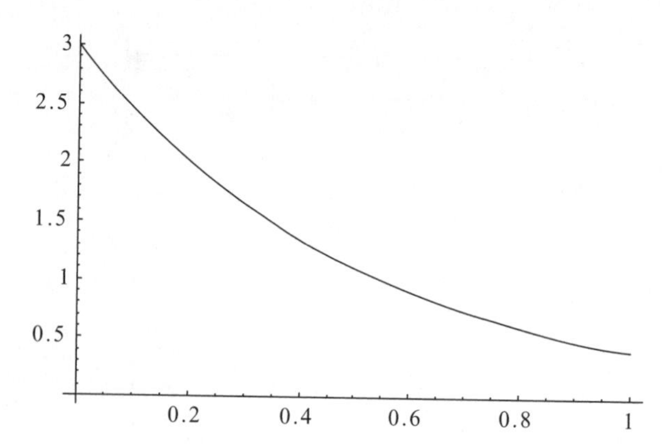

Fig. 6-16 A plot of an overdamped system, a decaying exponential.

If the roots of the characteristic equation are distinct and real, $s_1 \neq s_2$, then the solution is of the form

$$i(t) = Ae^{s_1 t} + Be^{s_2 t} \tag{6.29}$$

When these roots are both less than zero, we say that the system is *overdamped.* An example of an overdamped system is shown in Fig. 6-16; the current quickly decays to zero.

If the roots are real but $s_1 = s_2 = s$, the solution is of the form

$$i(t) = Ae^{st} + Bte^{st} \tag{6.30}$$

When $s < 0$, this is a *critically damped* system. In this case, the voltage or current rises smoothly to a peak value, then smoothly decays to zero. An example is shown in Fig. 6-17. Finally, if the roots are complex with $s_1 = a + ib$, $s_2 = a - ib$ the solution is of the form

$$i(t) = Ae^{at} \cos bt + Be^{at} \sin bt \tag{6.31}$$

When $a < 0$ we call this an underdamped system, which is a sinusoid with a decaying amplitude. An example is shown in Fig. 6-18. In all three cases, the constants A and B are determined from the initial conditions.

EXAMPLE 6-12

Find the zero-input voltage as a function of time across the capacitor in a series RLC circuit with $L = 2$, $C = 1/10$, and $R = 12$ and a voltage source $v(t) = 4 \cos 10t$. The initial conditions are $v(t) = 1$, $\dot{v}(t) = 0$.

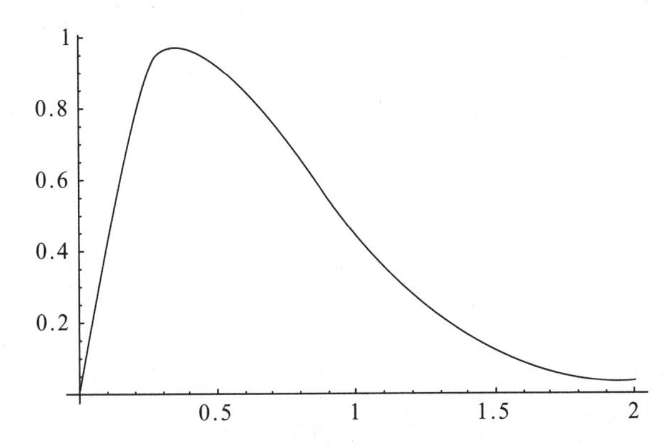

Fig. 6-17 A plot of a critically damped function, $f(t) = 8\,t\,e^{-3t}$.

SOLUTION

The differential equation to solve is

$$\frac{d^2 v_c}{dt^2} + 6\frac{dv_c}{dt} + 5v_c = 20\cos 10t$$

To find the zero-input solution, we solve the homogeneous equation

$$\frac{d^2 v_c}{dt^2} + 6\frac{dv_c}{dt} + 5v_c = 0$$

The characteristic equation is

$$s^2 + 6s + 5 = 0$$

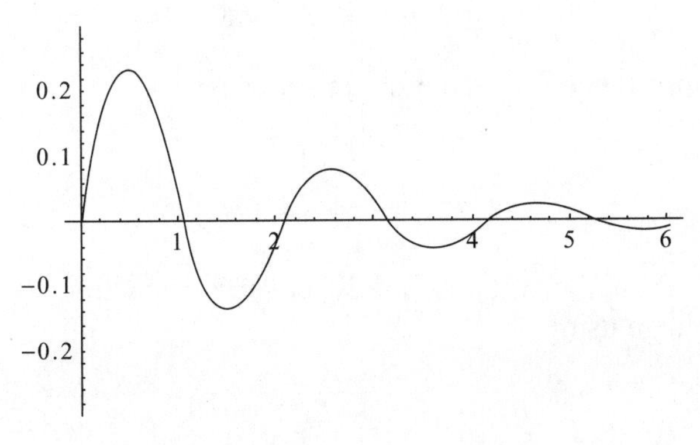

Fig. 6-18 An example of an underdamped function, $f(t) = 0.3e^{-0.5t}\sin 3t$.

The roots are

$$s_{1,2} = \frac{-6 \pm \sqrt{6^2 - 4(1)(5)}}{2(1)} = -1, -5$$

So the zero-input solution is

$$v_c(t) = Ae^{-t} + Be^{-5t}$$

If we were only interested in the zero-input solution, we could apply the initial conditions. Setting the voltage $v(0) = 1$ gives

$$A + B = 1$$

Setting the derivative to zero gives

$$0 = -A - 5B$$

We find $A = 5/4$, $B = -1/4$ so the zero-input solution is

$$v_c(t) = \frac{5}{4}e^{-t} - \frac{1}{4}e^{-5t}$$

Summary

In this chapter we considered dynamic circuit elements, the capacitor and the inductor. Capacitors in parallel add:

$$C_T = C_1 + C_2 + C_3 + \cdots$$

The total capacitance for capacitors in series is:

$$C_T = \frac{1}{1/C_1 + 1/C_2 + 1/C_3 + \cdots}$$

We measure capacitance C in Farads. Given a capacitor, we can relate the current and voltage using

$$i(t) = C\frac{dv}{dt} \quad \text{or} \quad v(t) = v(0) + \frac{1}{C}\int_0^t i(s)\,ds$$

In an inductor, this relation becomes

$$v = L\frac{di}{dt}$$

Where L is the inductance measured in Henries.

Quiz

1. A capacitor with $C = 1\ \mu$F is in an RC circuit with a 100 Ω resistor. What is the time constant?
2. How long does it take for the voltage in the circuit described in Problem 1 to decay to zero?
3. A current increases uniformly from 10 to 100 mA in 20 ms in a coil. This induces a voltage of 2 V. What is the inductance of the coil?
4. Consider an RL circuit with $R = 10\ \Omega$, $L = 2$ H in series with a voltage source with $v(t) = 24e^{-t}$. Find the total solution. Assume the initial current is zero.
5. Consider an RL circuit with $R = 8\ \Omega$, $L = 4$ H in series with a voltage source with $v(t) = 4\cos 10t$. Find the total solution. Assume the initial current is zero.
6. Consider an RL circuit with $R = 8\ \Omega$, $L = 4$ H in series with a voltage source with $v(t) = 4t$. Find the total solution. Assume the initial current is $i(t) = 2$ A.
7. The equation obeyed by the current in a parallel RLC circuit with zero input is

$$LC\frac{d^2i}{dt^2} + \frac{L}{R}\frac{di}{dt} + i = 0$$

Suppose that $L = 2$ H, $R = 4\ \Omega$, and $C = \frac{1}{2}$ F. If $i(0) = 1$, $v(0) = 0$, find $i(t)$.

CHAPTER 7

The Phasor Transform

In this chapter we consider sinusoidal sources and use the fact that they are closely related to complex numbers to develop a simplified method of calculation known as the *phasor transform.* Fortunately, sinusoidal sources are widely used. The household voltage in the United States is $v(t) = 170 \sin 377t$, so the topics covered in this chapter have wide applicability. Let's begin by reviewing some basic properties of complex numbers. If you are not familiar with complex numbers, don't despair, we include everything you need to know in this chapter.

Basics on Complex Numbers

In the sixteenth century mathematicians widened the scope of equations they could solve by introducing the notion of the square root of -1. Within the context of mathematics and physics, the square root of -1 is denoted by $i = \sqrt{-1}$. The symbol i is used because numbers of this type are sometimes known as *imaginary.* However, in electrical engineering it is traditional to denote the square root of -1 by

$$j = \sqrt{-1} \tag{7.1}$$

This is done, in part, because of tradition but mostly because we don't want to confuse $\sqrt{-1}$ with the current, which is denoted by the reserved symbol i. This way $4j$ has clear meaning, whereas if $4i$ was used in a set of equations describing a circuit its meaning would be ambiguous.

The square roots of a number $-a$ is given by $\sqrt{-a} = j\sqrt{a}$. For example

$$\sqrt{-4} = \sqrt{(-1)(4)} = \sqrt{-1}\sqrt{4} = j2$$

A *complex number* z is a number of the form

$$z = x + jy \tag{7.2}$$

We call x the *real part* of z and sometimes denote it by $x = \text{Re}(z)$. The *imaginary part* of z is $y = \text{Im}(z)$. Note that the real and imaginary parts of a complex number are real numbers. For example

$$z = 2 - 3j$$

In this case, $\text{Re}(z) = 2$ and $\text{Im}(z) = -3$.

Addition and subtraction of complex numbers proceeds as follows. If $z = x + jy$ and $w = u + jv$ we form their sum(difference) by adding(subtracting) their real and imaginary parts. That is

$$z \pm w = (x \pm u) + j(y \pm v) \tag{7.3}$$

To multiply two complex numbers, we write

$$\begin{aligned} zw &= (x + jy)(u + jv) = xu + jxv + jyu + j^2yv \\ &= (xu - yv) + j(xv + yu) \end{aligned} \tag{7.4}$$

The *complex conjugate* of a complex number z is denoted by $\bar{z}$ and is calculated by letting $j \to -j$. Hence, if $z = x + jy$, then

$$\bar{z} = x - jy \tag{7.5}$$

The complex conjugate is important because it can be used to denote the "length" of a complex number. We call this length the *magnitude*. You can think of a complex number as a vector in the $x - y$ plane pointing from the origin to the point (x, y), with the x component of the vector being the real part of z and the y component of the vector being the imaginary part of z. The magnitude is the product of the complex number and its conjugate

$$|z| = \sqrt{z\bar{z}} = \sqrt{(x + jy)(x - jy)} = \sqrt{x^2 + y^2} \tag{7.6}$$

Notice that this is the same number we would obtain for the magnitude of a vector $\vec{A} = x\hat{i} + y\hat{j}$.

To divide complex numbers, we use the conjugate in the following way. Again, let $z = x + jy$ and $w = u + jv$. Then

$$\frac{z}{w} = \frac{z\bar{w}}{z\bar{w}} = \frac{(x + jy)(u - jv)}{(u + jv)(u - jv)} = \frac{xu + yv + j(yu - xv)}{u^2 + v^2} \tag{7.7}$$

Polar Representation

So far we have been writing complex numbers using *Cartesian representation.* We can also write them in *polar form,* which is

$$z = re^{j\theta} \tag{7.8}$$

Here r is the magnitude of the complex number, again the length of a vector directed from the origin to the point (x, y). Therefore

$$r = \sqrt{z\bar{z}} \tag{7.9}$$

The phase or angle θ is the angle from the x axis to the vector representing the complex number. It is given by

$$\theta = \tan^{-1}(y/x) \tag{7.10}$$

In electrical engineering, it is common to denote the polar form (7.8) using a shorthand notation given by

$$z = r\angle\theta \tag{7.11}$$

Sinusoids and Complex Numbers

In many applications of circuit analysis, voltage and current sources are sinusoidal. For example, we can have $v(t) = V_0 \sin \omega t$ or $i(t) = I_0 \sin \omega t$. A useful

formula that you will become intimately familiar with if you study electrical engineering is *Euler's identity*. This allows us to relate the sine and cosine functions to complex exponentials. The cosine function is related to the exponential in the following way

$$\cos\theta = \frac{e^{j\theta} + e^{-j\theta}}{2} \tag{7.12}$$

And sine is related to the exponential via

$$\sin\theta = \frac{e^{j\theta} - e^{-j\theta}}{2j} \tag{7.13}$$

In addition, using (7.12) and (7.13) it is easy to show that

$$e^{\pm j\theta} = \cos\theta \pm j\sin\theta \tag{7.14}$$

These relations are easy to derive by writing down the Taylor series expansions of each term, but we won't worry about that and just accept them as given.

For particular angles, there are a few useful relationships that should be memorized. Using (7.14) notice that

$$e^{\pm j\pi} = -1 \tag{7.15}$$

This is sometimes written as the famous formula $e^{j\pi} + 1 = 0$. In addition note that

$$e^{\pm j\pi/2} = \pm j \tag{7.16}$$

It is also useful to note that

$$\frac{1}{j} = -j \tag{7.17}$$

EXAMPLE 7-1

Find the polar representation of $z = 4\sqrt{3} + 4j = 4(\sqrt{3} + j)$.

SOLUTION
First, looking in a table of trig functions, we note that

$$\cos\frac{\pi}{6} = \frac{\sqrt{3}}{2}, \quad \sin\frac{\pi}{6} = \frac{1}{2}$$

We have

$$z = 4(\sqrt{3} + j) = 8\left(\frac{\sqrt{3}}{2} + j\frac{1}{2}\right) = 8\left(\cos\frac{\pi}{6} + j\sin\frac{\pi}{6}\right)$$

Using Euler's identity (7.14) we see that the polar form of z is

$$z = 8\, e^{j\pi/6}$$

Hence, $r = 8$ and $\theta = \pi/6$, allowing us to write

$$z = 8\angle\pi/6$$

Multiplication and division of complex numbers in polar form is particularly easy. Let $z = re^{j\theta}$ and $w = \rho e^{j\phi}$. Then

$$zw = (re^{j\theta})(\rho e^{j\phi}) = r\rho e^{j(\theta+\phi)} \tag{7.18}$$

Hence, magnitudes multiply and angles add. We can write this in shorthand notation as

$$zw = r\rho\angle(\theta + \phi) \tag{7.19}$$

To divide two numbers in polar form, we divide the magnitudes and subtract the angles

$$\frac{z}{w} = \frac{re^{j\theta}}{\rho e^{j\phi}} = \frac{r}{\rho}e^{j(\theta-\phi)} \tag{7.20}$$

Or

$$\frac{z}{w} = \frac{r}{\rho}\angle(\theta - \phi) \tag{7.21}$$

Sinusoidal Sources

Now we consider in detail circuits with sinusoidal sources. A sinusoidal voltage source is one with the form

$$v(t) = V_0 \sin(\omega t + \phi) \tag{7.22}$$

Here V_0 is the amplitude (in volts) that gives the largest value that (7.22) can attain. We call ω the *radial frequency* with units rad/s and ϕ is the *phase angle.*

We can also have sinusoidally varying currents such as

$$i(t) = I_0 \sin(\omega t + \phi) \tag{7.23}$$

In this case the amplitude I_0 is measured in amps. The radial frequency is related to *frequency* by

$$\omega = 2\pi f \tag{7.24}$$

The units of f are *hertz* (Hz). The *period* T tells us the duration of a single cycle in the wave. It is related to frequency using

$$f = \frac{1}{T} \tag{7.25}$$

A sine wave $f(t) = 2 \sin t$ is shown in Fig. 7-1. Note that the amplitude gives the maximum value.

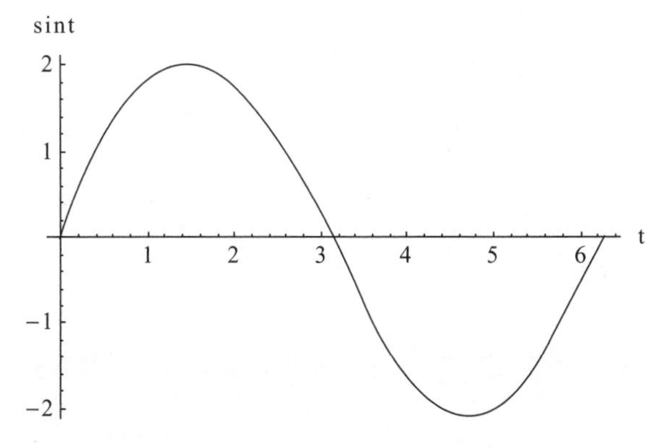

Fig. 7-1 A plot of $f(t) = 2 \sin t$.

Radians can be related to degrees by using

$$\theta(\text{radians}) = \frac{\pi}{180^\circ} \times \theta\,(\text{degrees})$$
$$\theta(\text{degrees}) = \frac{180^\circ}{\pi} \times \theta\,(\text{radians}) \tag{7.26}$$

Leading and Lagging

In electrical engineering you often hear the terms *leading* and *lagging*. Let $v_1(t) = V_0 \sin(\omega t + \phi)$ and $v_2(t) = V_0 \sin(\omega t)$. Both voltages have the same radial frequency, but we say that $v_1(t)$ leads $v_2(t)$ by ϕ radians or degrees (depending on the units used). This means that the features in the waveform $v_1(t)$ appear earlier in time than the features in $v_2(t)$. Otherwise they are the same waveforms. Consider

$$v_1(t) = 170 \sin(377t + 20^\circ), \quad v_2(t) = 170 \sin(377t)$$

In this case $v_1(t)$ leads $v_2(t)$ by 20°. Alternatively, we can say that $v_2(t)$ lags $v_1(t)$ by 20°.

In Fig. 7-2, we show a plot of $f(t) = 2 \sin t$ together with $g(t) = 2 \sin(t + \pi/6)$. The dashed line is $f(t) = 2 \sin t$, which lags $g(t) = 2 \sin(t + \pi/6)$ because the features of $g(t)$ appear *earlier*.

Now suppose that $g(t) = 2 \sin(t - \pi/6)$. This wave lags $f(t) = 2 \sin t$, meaning that its features appear later in time. This is illustrated in Fig. 7-3.

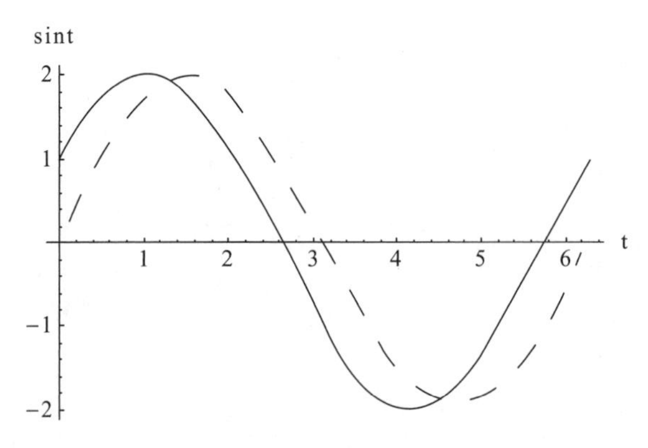

Fig. 7-2 The wave $g(t) = 2 \sin(t + \pi/6)$ leads $f(t) = 2 \sin t$, its features appear earlier in time.

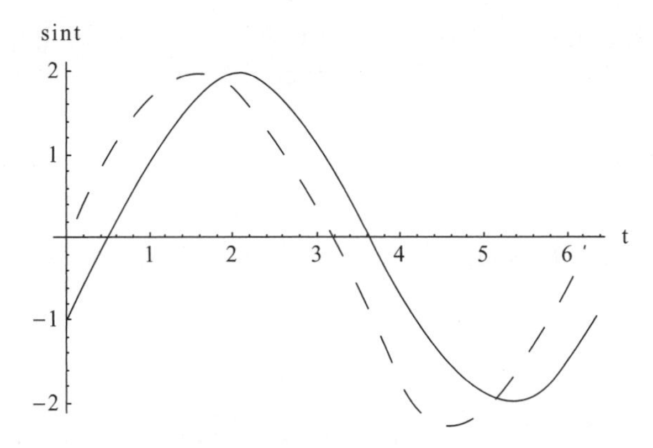

Fig. 7-3 The wave $g(t) = 2\sin(t - \pi/6)$ (dashed line) *lags* $f(t) = 2\sin t$ (solid line), now $f(t) = 2\sin t$ appears earlier in time.

If two sinusoidal waveforms have a 0° phase difference, we say that they are *in phase.* If the phase difference is 90°, we say that the waves are 90° *out of phase.*

Effective or RMS Values

The *effective* or *RMS* value of a periodic signal is the positive dc voltage or current that results in the same power loss in a resistor over one period. If the current or voltage is sinusoidal, we divide the amplitude by $\sqrt{2}$ to get the RMS value. That is

$$V_{\text{eff}} = \frac{V_0}{\sqrt{2}}, \quad I_{\text{eff}} = \frac{I_0}{\sqrt{2}} \tag{7.27}$$

Dynamic Elements and Sinusoidal Sources

Suppose that a voltage $v(t) = V\sin(\omega t + \phi)$ is across a capacitor C. The current in the capacitor is

$$i(t) = C\frac{dv}{dt} = \omega CV\cos(\omega t + \phi) \tag{7.28}$$

The amplitude and hence the maximum value attained by the current is $I = \omega CV$. Rewriting this in terms of the voltage we have

$$V = \frac{1}{\omega C} I$$

Notice that this resembles Ohm's law. So we can denote a resistance by

$$X_C = -\frac{1}{\omega C} \tag{7.29}$$

We call this quantity the *capacitive reactance.* The negative sign results from the phase shift that occurs relating voltage to current.

Now consider an inductor carrying a current $i(t) = I \sin(\omega t + \phi)$. The voltage across the inductor is given by

$$v(t) = L\frac{di}{dt} = \omega L I \cos(\omega t + \phi)$$

Following the same logic used when considering a capacitor, we note that the maximum voltage in the inductor is $V = \omega L I$. Once again this is an Ohm's law type relation with resistance $R = \omega L$. We call this the inductive reactance

$$X_L = \omega L \tag{7.30}$$

Notice that the inductive reactance and capacitive reactance have the same units as resistance. Unlike a resistor, however, the resistance in an inductor or capacitor is frequency dependent. As the frequency increases

- The resistance of a capacitor *decreases.*
- The resistance of an inductor *increases.*

The Phasor Transform

A *phasor* is a complex representation of a phase-shifted sine wave. If

$$f(t) = A \cos(\omega t + \phi) \tag{7.31}$$

Then the phasor is given by

$$\mathbf{F} = A\angle\phi \tag{7.32}$$

To see how this works, we begin by considering Euler's identity (7.14). Since $e^{j(\omega t+\phi)} = \cos(\omega t + \phi) + j\sin(\omega t + \phi)$, we can take (7.31) to be the real part of this expression, that is

$$f(t) = \mathrm{Re}[Ae^{j(\omega t+\phi)}] \tag{7.33}$$

If a source is given as a sine wave, we can always rewrite it as a cosine wave because

$$\cos(x - 90^\circ) = \sin x \tag{7.34}$$

In a given circuit with a source $f(t) = A\cos(\omega t + \phi)$, the frequency part will be the same for all components in the circuit. Hence, we can do our analysis by focusing on the phase lead or lag for each voltage and current in the circuit. This is done by writing each voltage and current by using what is called a *phasor transform*. We denote phasor transforms with boldface letters. For $f(t) = A\cos(\omega t + \phi)$, the phasor transform is just

$$\mathbf{F} = Ae^{j\phi} \tag{7.35}$$

The functions $f(t) = A\cos(\omega t + \phi)$ and $\mathbf{F} = Ae^{j\phi}$ constitute a *phasor transform pair.* We write this relationship as

$$f(t) \Leftrightarrow \mathbf{F} \tag{7.36}$$

EXAMPLE 7-2

If $i(t) = 20\cos(12t + 30^\circ)$, what is the phasor transform?

SOLUTION

First note that the current $i(t)$ is the real part of

$$i(t) = \mathrm{Re}[20e^{j(12t+30^\circ)}] = \mathrm{Re}[20\cos(12t + 30^\circ) + j20\sin(12t + 30^\circ)]$$

The phasor transform is

$$\mathbf{I} = 20e^{j30^\circ}$$

Or we can write it in shorthand as

$$\mathbf{I} = 20\angle 30^\circ$$

Properties of the Phasor Transform

Phasor transforms are very useful in electrical engineering because they allow us to convert differential equations into algebra. In particular, the differentiation operation is converted into simple multiplication. Again, let's start with $f(t) = A\cos(\omega t + \phi)$, which allows us to write

$$f(t) = \text{Re}\,[Ae^{j(\omega t+\phi)}]$$

Then

$$\frac{df}{dt} = -\omega A \sin(\omega t + \phi)$$

But notice that

$$\frac{d}{dt} Ae^{j(\omega t+\phi)} = j\omega Ae^{j(\omega t+\phi)}$$

Since $f(t) = \text{Re}\,[Ae^{j(\omega t+\phi)}]$, it follows that

$$\frac{df}{dt} = j\omega \text{Re}\,[Ae^{j(\omega t+\phi)}] \tag{7.37}$$

For the phasor transform, we have the relation

$$\frac{df}{dt} \Leftrightarrow j\omega Ae^{j\phi} \tag{7.38}$$

Now let's consider integration. We integrate from the time 0 just before the circuit is excited to some time t, so define

$$g(t) = \int_{0^-}^{t} Ae^{j(\omega\tau+\phi)} d\tau$$

Noting that at $t = 0^-$, we take the function to be zero. Letting $u = j(\omega\tau + \phi)$ we have

$$du = j\omega d\tau, \Rightarrow d\tau = \frac{1}{j\omega} du$$

And we obtain

$$g(t) = \frac{1}{j\omega} \int Ae^{ju} du = \frac{1}{j\omega} Ae^{j(\omega\tau+\phi)}\Big|_{0^-}^{t} = \frac{1}{j\omega} Ae^{j(\omega t+\phi)}$$

Hence, integration, which is the inverse operation to differentiation, results in division by $j\omega$ in the phasor domain. Given a sinusoidal function $f(t)$ with phasor transform $\mathbf{F}$ we have the phasor transform pair

$$\int_{0^-}^{t} f(\tau)\,d\tau \Leftrightarrow \frac{1}{j\omega}\mathbf{F} \tag{7.39}$$

Circuit Analysis Using Phasors

With differentiation and integration turned into simple arithmetic we can do steady-state analysis of sinusoidally excited circuits quite easily. The current flowing through a capacitor is given by

$$i(t) = C\frac{dv}{dt}$$

When we work with phasors, this relation becomes

$$\mathbf{I} = j\omega C\mathbf{V} \tag{7.40}$$

The voltage across a capacitor is related to the current via

$$v(t) = \frac{1}{C}\int_{0^-}^{t} i(\tau)\,d\tau$$

The phasor transform of this relation is

$$\mathbf{V} = \frac{1}{j\omega C}\mathbf{I} \tag{7.41}$$

Now let's turn to the inductor. The voltage across an inductor is related to the current through the time derivative

$$v(t) = L\frac{di}{dt}$$

Hence, the phasor relationship is

$$\mathbf{V} = j\omega L\mathbf{I} \tag{7.42}$$

Finally, the current flowing through an inductor is related to the voltage using the integral

$$i(t) = \frac{1}{L}\int_{0^-}^{t} v(\tau)\,d\tau$$

Therefore, the phasor transform gives the following relationship

$$\mathbf{I} = \frac{1}{j\omega L}\mathbf{V} \tag{7.43}$$

We refer to these quantities as *admittances*. Given these relations, our approach to doing steady-state analysis of sinusoidal circuits will be as follows

- Compute the phasor transform of each quantity in the circuit.
- Solve for the unknown phasor currents and voltages algebraically.
- Transform back to write down the unknowns as functions of time.

We proceed with some examples.

EXAMPLE 7-3

Find the steady-state current flowing through the capacitor in the circuit shown in Fig. 7-4. Take the voltage source to be $v(t) = 20\cos 100t$, $R = 1/4$ and $C = 1/10$.

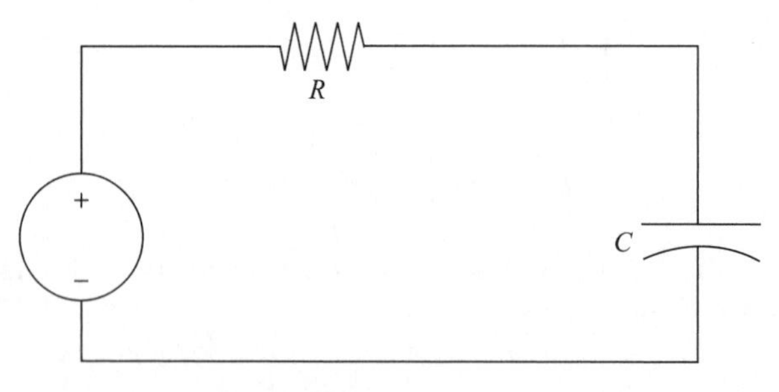

Fig. 7-4 Circuit for Example 7-3. The voltage source is $v(t) = 20\cos 100t$, and $R = 1/4$ and $C = 1/10$.

SOLUTION

We can solve the circuit by using KVL. We obtain

$$-20\cos 100t + Ri(t) + \frac{1}{C}\int_0^t i(\tau)\,d\tau = 0$$

Rather than try to solve this integral equation, we can transform to the phasor domain. First, the voltage source has zero phase and so transforms as

$$v(t) = 20\cos 100t \Leftrightarrow 20\angle 0^\circ$$

The integral becomes

$$\frac{1}{C}\int_0^t i(\tau)\,d\tau \Leftrightarrow \frac{1}{j\omega C}\mathbf{I} = \frac{1}{j(100)(1/10)}\mathbf{I} = \frac{1}{j10}\mathbf{I} = -j\frac{1}{10}\mathbf{I}$$

where we used (7.17). With these transforms in hand we can rewrite the KVL equation as

$$-j\frac{1}{10}\mathbf{I} + \frac{1}{4}\mathbf{I} - 20\angle 0^\circ = 0, \Rightarrow$$

$$\mathbf{I}\left(\frac{1}{4} - j\frac{1}{10}\right) = 20\angle 0^\circ$$

To solve for the current, we need to convert the complex number on the left into polar notation. First, we calculate the magnitude

$$r = \sqrt{\left(\frac{1}{4}\right)^2 + \left(\frac{1}{10}\right)^2} = 0.27$$

The phase angle is

$$\theta = \tan^{-1}\left(\frac{-1/10}{1/4}\right) = -0.4^\circ$$

Hence, we have

$$\mathbf{I}(0.27\angle -0.4^\circ) = 20\angle 0^\circ, \Rightarrow$$

$$\mathbf{I} = \frac{20\angle 0^\circ}{0.27\angle -0.4^\circ} = 74\angle 0^\circ - (-0.4^\circ) = 74\angle 0.4^\circ$$

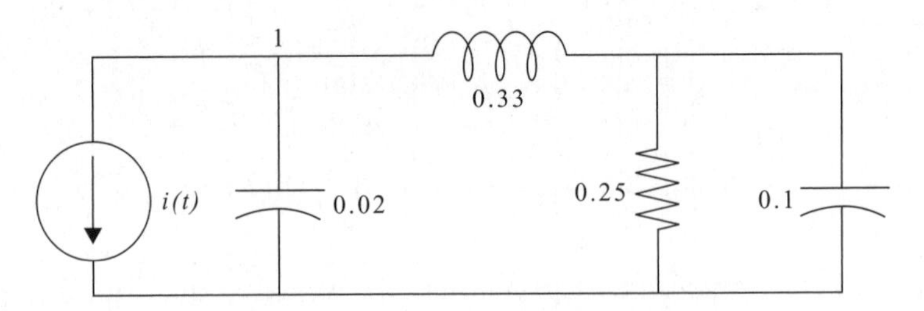

Fig. 7-5 Circuit for Example 7-4.

Inverting, we find that the current through the capacitor is

$$i(t) = 74\cos(100t + 0.4^\circ)$$

In our next example, consider the circuit in Fig. 7-5.

EXAMPLE 7-4
If $i(t) = 30\cos(100t - 60^\circ)$, find the steady-state voltage across the $C = 0.1$ capacitor.

SOLUTION
The phasor transform of the input current is

$$i(t) = 30\cos(100t - 60^\circ) \Leftrightarrow 30\angle{-60^\circ}$$

For the $C = 0.2$ capacitor, with $\omega = 100$ the admittance is $j2$. The admittance for the inductor is $j0.33$, the admittance for the resistor is $1/R = 4$, and the admittance for the final capacitor on the right is $j1$. Let us denote the voltage drop at node 1 in Fig. 7-5 by $v_1(t)$ and the voltage drop across the $C = 0.1$ capacitor as $v_2(t)$.

Using the admittances, we can solve the circuit. Applying KCL at node 1 as labeled in Fig. 7-5, using + for currents leaving the node, we have

$$30\angle{-60^\circ} + j2\mathbf{V}_1 + (-j0.33)(\mathbf{V}_1 - \mathbf{V}_2)$$
$$\Rightarrow$$
$$j(1.67)\mathbf{V}_1 + j0.33\mathbf{V}_2 = -30\angle{-60^\circ} = 30\angle 120^\circ$$

Now let's apply KCL to the node just above the $C = 0.1$ capacitor. We have

$$j\mathbf{V}_2 + 4\mathbf{V}_2 - j0.33(\mathbf{V}_2 - \mathbf{V}_1) = 0$$
$$\Rightarrow$$
$$j0.33\mathbf{V}_1 + (4 + j0.67)\mathbf{V}_2 = 0$$

We now have two equations and two unknowns for this circuit

$$j(1.67)\mathbf{V}_1 + j0.33\mathbf{V}_2 = 30\angle 120^\circ$$
$$j0.33\mathbf{V}_1 + (4 + j0.67)\mathbf{V}_2 = 0$$

Eliminating $\mathbf{V}_1$ we arrive at the solution in phasor space for $\mathbf{V}_2$

$$\mathbf{V}_2 = 0.53 - j1.36$$

The magnitude of this complex number is

$$r = \sqrt{(0.53)^2 + (-1.36)^2} = 1.46$$

The phase angle is

$$\phi = \tan^{-1}\left(\frac{-1.36}{0.53}\right) = -68.71^\circ$$

Hence

$$\mathbf{V}_2 = 1.46\angle -68.71^\circ$$

To find the voltage as a function of time, we recall that the frequency stays the same as the frequency given for the input. Therefore, we have the voltage across the capacitor as a function of time as

$$v_2(t) = 1.46\cos(100t - 68.71^\circ)$$

Impedance

We define the *impedance* as the ratio of voltage to current for a given circuit element. This is simply a generalization of Ohm's law where

$$Z = \frac{V}{I} \tag{7.44}$$

EXAMPLE 7-5

Find the Thevinin equivalent of the circuit illustrated in Fig. 7-6 where $v(t) = 120\sin 377t$. The impedance of the circuit element Z is $Z = 4 - j2$.

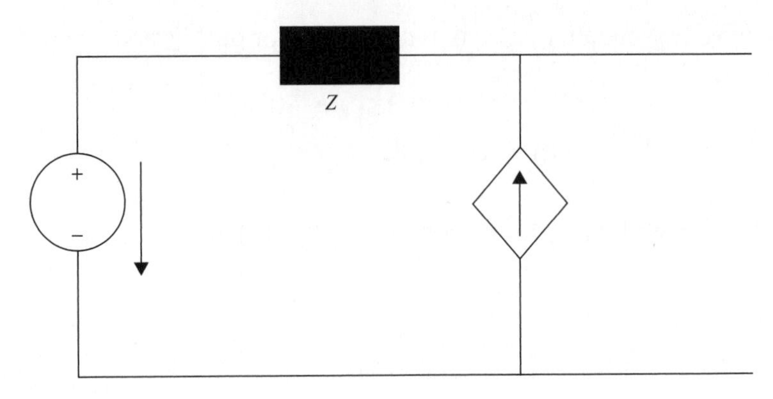

Fig. 7-6 In Example 7-5, we find the Thevenin equivalent circuit by using phasor transforms.

The dependent current source is $i = 0.8i_s(t)$, where i_s is the current flowing through the voltage source.

SOLUTION

We attach a current source across the open terminals as shown in Fig. 7-7.

We denote the current flowing through the voltage source as $i_s(t)$, the current source added to the terminals on the right in Fig. 7-7 by $i_0(t)$, and the voltage across this current source by $v_0(t)$.

The phasor transform of the source voltage $v(t) = 120 \sin 377t$ is $\mathbf{V}_s = 120\angle 0°$. Applying KVL to the outside loop in Figure 7-7 leads to

$$-\mathbf{V}_0 + (4 - j2)(0.8\mathbf{I}_s + \mathbf{I}_0 + 120\angle 0°)$$

Solving for $\mathbf{V}_0$ we have

$$\mathbf{V}_0 = (4 - j2)(0.8\mathbf{I}_s + \mathbf{I}_0 + 120\angle 0°)$$

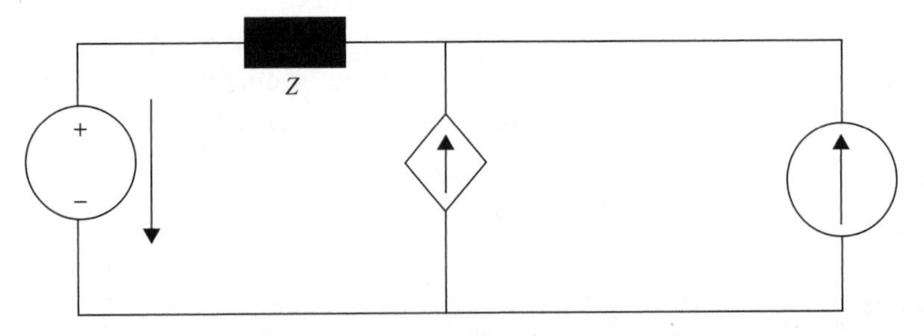

Fig. 7-7 To find the Thevenin equivalent circuit, we attach a current source to the open terminals of the circuit shown in Fig. 7-6.

But KCL (+ for currents entering) at the node above the dependent current source gives

$$-\mathbf{I}_s + 0.8\mathbf{I}_s + \mathbf{I}_0 = 0$$
$$\Rightarrow \mathbf{I}_s = 5\mathbf{I}_0$$

So we have

$$\mathbf{V}_0 = (4 - j2)(0.8(5)\mathbf{I}_0 + \mathbf{I}_0 + 120\angle 0^\circ)$$

Therefore we have the following equation

$$\mathbf{V}_0 = (20 - j10)\mathbf{I}_0 + 120\angle 0^\circ$$

The Thevenin equivalent impedance is therefore

$$\mathbf{Z}_{\text{TH}} = 20 - j10$$

While the Thevenin equivalent voltage is then

$$\mathbf{V}_{\text{TH}} = 120\angle 0^\circ$$

The Thevenin equivalent circuit is shown in Fig. 7-8.

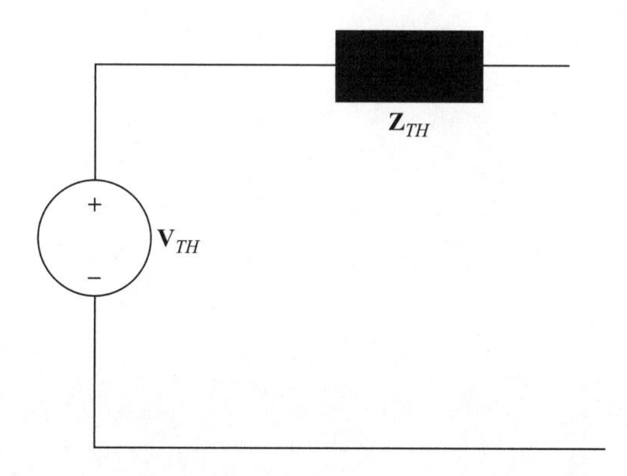

Fig. 7-8 The Thevenin equivalent circuit for Example 7-5.

Summary

In this chapter we considered the use of the phasor transform, which simplifies steady-state sinusoidal analysis. Using the fact that $e^{j\theta} = \cos\theta + j\sin\theta$ we can simplify analysis by using the polar representation of a complex number

$$z = re^{j\theta}$$

which is often written using a shorthand notation $z = r\angle\theta$ in electrical engineering. For a voltage source given by $v(t) = V_0 \sin(\omega t + \phi)$, V_0 is the amplitude (in volts) which gives the largest value that the voltage can attain. We call ω the radial frequency with units rad/s and ϕ is the phase angle. To compute the phasor transform of a time varying function $f(t) = A\cos(\omega t + \phi)$, we write

$$f(t) = \text{Re}[Ae^{j(\omega t+\phi)}]$$

We then solve the circuit by working with the simpler quantity $\mathbf{F} = Ae^{j\phi}$, which allows us to turn differentiation into multiplication and integration into division using

$$\frac{df}{dt} \Leftrightarrow j\omega Ae^{j\phi} \quad \text{and} \quad \int_{0^-}^{t} f(\tau)\,d\tau \Leftrightarrow \frac{1}{j\omega}\mathbf{F}$$

When analyzing sinusoidal circuits with inductors and capacitors, two quantities that are useful when characterizing the circuit are the capacitive reactance

$$X_C = -\frac{1}{\omega C}$$

and the inductive reactance

$$X_L = \omega L$$

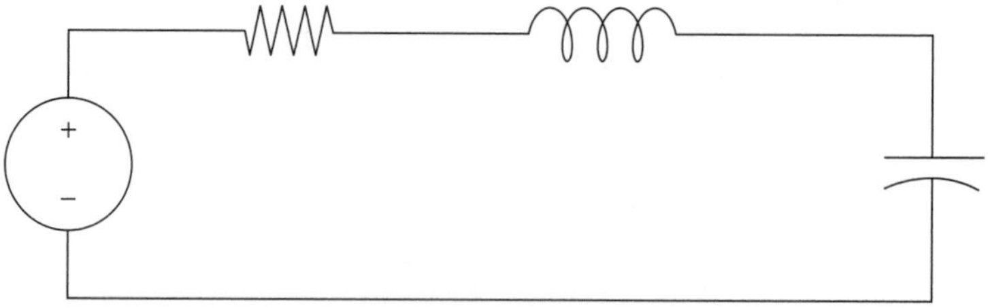

Fig. 7-9 Circuit for Problem 5.

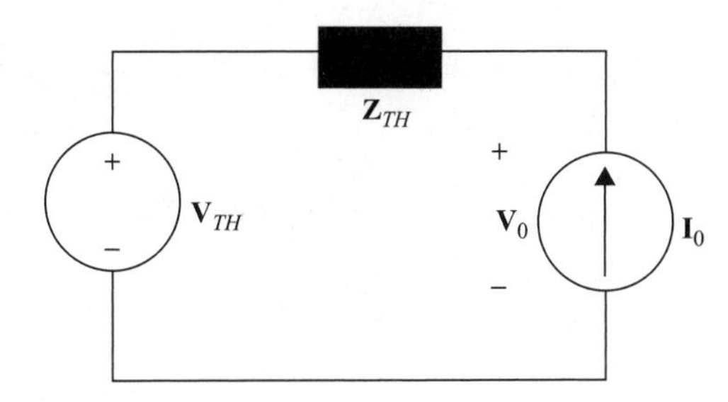

Fig. 7-10 This circuit gives the relation used to deduce the Thevenin equivalent impedance and voltage as used in Example 7-5.

Quiz

1. Write $z = \frac{3}{2} - j\frac{3\sqrt{3}}{2}$ in polar form.
2. Let $i_1(t) = 12 \sin 20t,\ i_2(t) = 12 \sin(20t + 10°)$. What is the amplitude of each wave? Are the waves in phase?
3. The voltage across a resistor is $v(t) = V \sin \omega t$. What is the average power?
4. Define impedance.
5. Find the current flowing through the circuit shown in Fig. 7-9 by using phasor transform analysis. The source voltage is $v(t) = 40\sqrt{2} \sin(4t + 20°)$, $R = 6$, $L = 2$, $C = 1/16$.
6. Derive the relation for the Thevenin equivalent circuit and applied current source by applying KVL to the circuit shown in Fig. 7-10.

CHAPTER 8

Frequency Response

By using capacitors and inductors in combination, we can construct circuits that will respond to a specific frequency of our choosing. To see how such a circuit works, we consider a simple series LC circuit, shown in Fig. 8-1.

Natural Frequencies

We consider the series circuit shown in Fig. 8-1. Let's suppose that the voltage source is sinusoidal such that

$$v_s(t) = V_0 \cos \omega t \tag{8.1}$$

Although we have seen circuits like this before, let's go through the solution process to reinforce what we've learned about circuits. The first item to notice about Fig. 8-1 is that the same current flows through each circuit element. Using the fact that

$$i_C = C\frac{dv_C}{dt}, \quad v_L = L\frac{di_L}{dt}$$

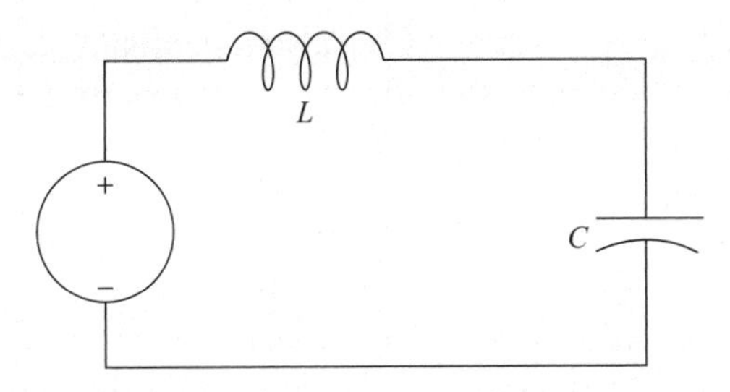

Fig. 8-1 A circuit with a voltage source $v_s(t)$, inductor L, and capacitor C in series.

and taking $i = i_C = i_L$, we have

$$v_L = L\frac{di}{dt} = LC\frac{d^2v_C}{dt^2}$$

Therefore KVL going in a clockwise loop around the circuit in Fig. 8-1 gives

$$LC\frac{d^2v_C}{dt^2} + v_C = V_0 \cos \omega t \tag{8.2}$$

Following the procedure used in Chapter 6, let's consider the zero-input response first. The equation for the zero-input response is found by setting $v_s = 0$ giving

$$LC\frac{d^2v_C}{dt^2} + v_C = 0$$

Using standard techniques of differential equations, we find the characteristic equation to be

$$s^2 + \frac{1}{LC} = 0$$

The complex roots are

$$s = \pm j\frac{1}{\sqrt{LC}}$$

With complex roots, we will have a purely sinusoidal solution. Therefore, we identify the complex roots with the *natural frequency* of the circuit

$$\omega_0 = \frac{1}{\sqrt{LC}}$$

We call this the natural frequency because this is the frequency without any sources—the zero-input response. So the zero-input or homogeneous solution of (8.2) is given by

$$v_C^H(t) = c_1 \cos(\omega_0 t + \phi) \tag{8.3}$$

where c_1 is a constant to be determined by initial conditions.

To determine the particular solution, we assume a solution of the form

$$v_C^P(t) = c_2 \cos(\omega t + \theta)$$

Hence

$$\frac{d^2 v_C^P}{dt^2} = -\omega^2 c_2 \cos(\omega t + \theta)$$

Therefore, inserting the particular solution into the complete differential equation (8.2) gives

$$-\omega^2 LC c_2 \cos(\omega t + \theta) + c_2 \cos(\omega t + \theta) = V_0 \cos \omega t$$

Cleaning up a bit we have

$$c_2 \cos(\omega t + \theta)(1 - \omega^2 LC) = V_0 \cos \omega t$$

Using the trig identity for the cosine of a sum of two arguments we find that

$$\cos(\omega t + \theta) = \cos \omega t \cos \theta - \sin \omega t \sin \theta$$

So we can write

$$c_2(\cos \omega t \cos \theta - \sin \omega t \sin \theta) = \frac{V_0}{1 - \omega^2 LC} \cos \omega t$$

There are no sine terms on the right-hand side, so we have $\sin \omega t \sin \theta = 0$. This will be true if we fix $\theta = 0$ and so $\cos \theta = 1$ and we are left with the

matching condition

$$c_2 \cos \omega t = \frac{V_0}{1 - \omega^2 LC} \cos \omega t$$

This tells us to take the coefficient as

$$c_2 = \frac{V_0}{1 - \omega^2 LC}$$

And we obtain

$$v_C^P(t) = \frac{V_0}{1 - \omega^2 LC} \cos \omega t$$

for the particular solution. The total solution is the sum of the homogeneous and particular solutions giving

$$v_C(t) = v_C^H(t) + v_C^P(t) = c_1 \cos(\omega_0 t + \phi) + \frac{V_0}{1 - \omega^2 LC} \cos \omega t$$

At this point, the zero-input response really isn't of much interest to us. We want to see how the circuit responds to inputs of different frequency. Note that the coefficient of the particular solution actually contains the natural frequency in the denominator

$$\frac{V_0}{1 - \omega^2 LC} = \frac{V_0}{LC\left(\dfrac{1}{LC} - \omega^2\right)} = \frac{V_0}{LC\left(\omega_0^2 - \omega^2\right)}$$

The particular solution will go to infinity or blow up if

$$\omega_0^2 - \omega^2 = 0$$

That is, if the frequency of the source voltage or input exactly matches the natural frequency of the circuit

$$\omega = \omega_0 \tag{8.4}$$

This frequency-matching condition tells us that the voltage across the capacitor (and hence the current in the circuit) will blow up if the input frequency matches the natural frequency. For this reason, we call this the *resonant frequency* of the circuit.

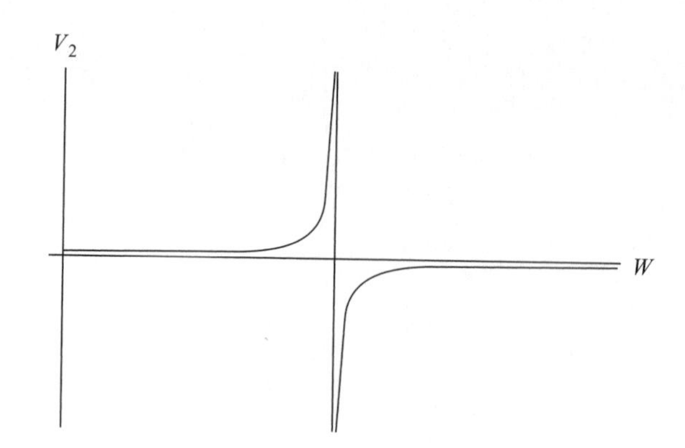

Fig. 8-2 A plot of the ratio of the output voltage to the input voltage for the series LC circuit shown in Fig. 8-1.

Considering only the zero-input response of the circuit, the ratio of the output voltage, which we denote $v_r(t)$ for response to the source or input voltage $v_s(t)$, gives us the coefficient

$$\frac{1}{1 - \omega^2 LC}$$

A plot of this function against frequency is shown in Fig. 8-2. The point at which the function blows up is given by (8.4).

In our case, we are considering *ideal* inductors and capacitors, so the ratio blows up to infinity. In real life, inductors and capacitors are not ideal, meaning that they have losses. As a result the ratio of output to input voltage does not blow up to infinity, but instead just gets *very large*. In a real circuit, a plot of the ratio of output to input voltage might look something like that shown in Fig. 8-3. The voltage attains a peak value at the natural or resonant frequency, which is much larger than the response at other frequencies. However, the response does not blow up or go to infinity. In many cases, it is desired to design a circuit in this way so that it only responds for a specific frequency. For example, we may want to tune into a particular radio frequency. We could do this by having a variable capacitor and set C so that the natural frequency of the circuit matched the frequency of the input signal from the radio transmitter.

The Frequency Response of a Circuit

In this section we investigate the frequency response of a circuit in more detail by using phasor analysis. We are interested in characterizing the behavior of the

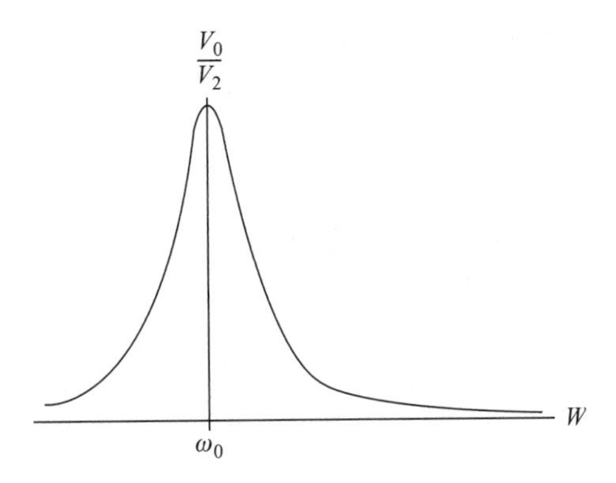

Fig. 8-3 In a real circuit, the response at the natural frequency will be much larger, but not infinite.

circuit for different frequencies by examining its impedance. For example, we may want to know not only the peak or resonant frequency of the circuit, but also over what range of frequencies the circuit is inductive or capacitive. First, let's review some basic concepts of AC circuit analysis we began to discuss in Chapter 6. The impedance of a circuit is

$$Z = R + jX \tag{8.5}$$

where R is the resistance of the circuit and X is the reactance of the circuit. For an inductor, the reactance is positive and takes the form

$$X_L = \omega L \tag{8.6}$$

Notice that the reactance for an inductive circuit depends linearly on frequency and is increasing with frequency. The reactance of a capacitor is

$$X_C = -\frac{1}{\omega C} \tag{8.7}$$

Notice, as the frequency increases, capacitive reactance decreases. In polar form the impedance is

$$\mathbf{Z} = \sqrt{R^2 + X^2} \angle \tan^{-1}\left(\frac{X}{R}\right) \tag{8.8}$$

For impedances, Ohm's law takes the form

$$\mathbf{V} = \mathbf{ZI} \tag{8.9}$$

The inverse of impedance is *admittance*

$$\mathbf{Y} = \frac{1}{\mathbf{Z}} \tag{8.10}$$

Admittance can be written in terms of real and imaginary parts as

$$Y = G + jB \tag{8.11}$$

We call G the *conductance* of the circuit and B the *susceptance*. In polar form

$$\mathbf{Y} = \sqrt{G^2 + B^2} \angle \tan^{-1}\left(\frac{B}{G}\right) \tag{8.12}$$

By examining the response of a circuit that may be a current or a voltage, we can determine whether the circuit is inductive or capacitive.

EXAMPLE 8-1

A load has a voltage $\mathbf{V} = 120\angle 25^\circ$ and current $\mathbf{I} = 60\angle 60^\circ$. Find the impedance and determine a series circuit that will model the load. Is the circuit inductive or capacitive? Assume that $\omega = 377$ rad/s.

SOLUTION

Using Ohm's law (8.9) we find

$$\mathbf{Z} = \frac{\mathbf{V}}{\mathbf{I}} = \frac{120\angle 25^\circ}{60\angle 60^\circ} = 2\angle -35^\circ$$

We can write the impedance in the form $Z = R + jX$ by using the relations

$$R = |Z| \cos\phi, \quad X = |Z| \sin\phi \tag{8.13}$$

For the impedance with the given parameters in this problem, we find

$$Z = 1.64 - j1.15$$

Now, notice from (8.7) that the capacitive reactance is negative. This tells us that, if we calculate the impedance for a given circuit and find the reactance is negative, the circuit is capacitive. Hence, we determined that the load can be modeled by a resistor and a capacitor in series. The resistor is just

$$R = 1.64\ \Omega$$

To find the capacitance, we use

$$X = -\frac{1}{\omega C}, \Rightarrow C = \frac{1}{(377)(1.15)} = 2.3 \text{ mF}$$

We can characterize the response of a circuit in the following way. Let **E** be a phasor that represents a source or *excitation* of the circuit. It may be a voltage or a current. Now let **R** be the response of the circuit. Then

$$\mathbf{R} = \mathbf{HE} \tag{8.14}$$

where **H** is the phasor network or *transfer function.* If the relationship is of the form

$$\mathbf{V} = \mathbf{ZI}$$

then we say that **H** is a *transfer impedance.* If the relationship is of the form

$$\mathbf{I} = \mathbf{YV}$$

then we say that **H** is a transfer admittance. We can determine the resonant frequencies of a given circuit by considering the transfer function. Specifically we write it in polar form

$$\mathbf{H} = |H| \angle\theta \tag{8.15}$$

The condition for resonance is met when the phase angle of the transfer function vanishes, that is,

$$\theta = 0 \tag{8.16}$$

for (8.15). To see why this is the case, we note that resonance occurs when the response is in phase with the excitation. Remember that when we multiply two phasors together, as we do in (8.14), we multiply the magnitudes and add up the phase angles. Looking at (8.14)

$$\angle\theta_R = \angle\theta_H + \angle\theta_E \tag{8.17}$$

To have the condition where the phase angle of **R** is the same as the phase angle of **E,** the phase angle of **H** must be zero. Another way to look at this is to

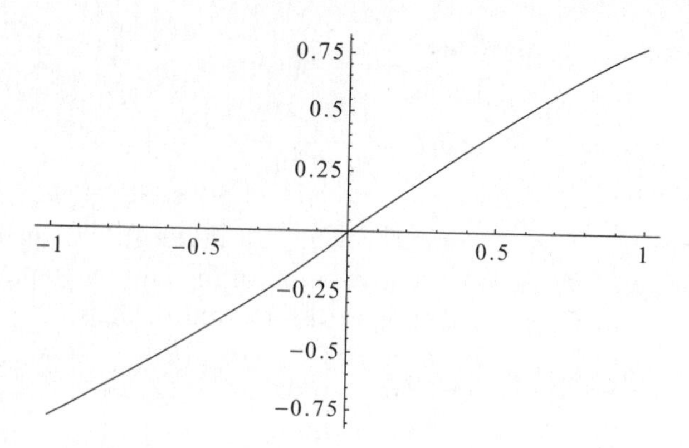

Fig. 8-4 A plot of the arctangent function.

simply examine a plot of the arctangent function shown in Fig. 8-4. It goes to zero when the argument is zero, hence if

$$H = f(\omega) + jg(\omega)$$

Then $\theta_H = 0$ when the imaginary part of the transfer function is zero because

$$\theta_H = \tan^{-1}\left(\frac{g}{f}\right) = 0, \Rightarrow g = 0$$

In the next example, we see how the method can determine the resonant frequency with a series RLC circuit, as shown in Fig. 8-5.

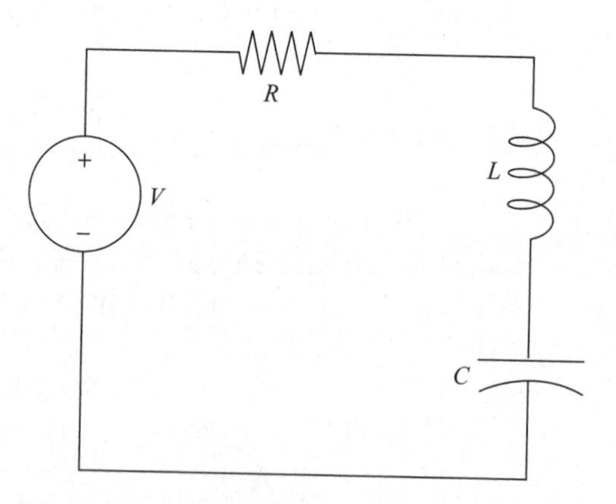

Fig. 8-5 We find the resonant frequency for a series RLC circuit connected to a sinusoidal voltage source.

EXAMPLE 8-2

Find the resonant frequencies for a series RLC circuit as shown in Fig. 8-5, when $v_s(t) = V_0 \cos \omega t$. The response of the circuit is the current flowing through the capacitor.

SOLUTION

With one loop its easy to apply KVL

$$V_0 \cos \omega t = Ri(t) + L\frac{di}{dt} + v_c$$

Now we recall that $i_c = C dv_c/dt$ and this becomes

$$V_0 \cos \omega t = RC\frac{dv_c}{dt} + LC\frac{d^2 v_c}{dt^2} + v_c$$

Taking the phasor transform of this equation where $d/dt \to j\omega$ gives

$$\begin{aligned} V_0\angle 0^\circ &= j\omega RC\mathbf{V_C} - \omega^2 LC\mathbf{V_C} + \mathbf{V_C} \\ &= \mathbf{V_C}(1 - \omega^2 LC + j\omega RC) \end{aligned}$$

So we find that

$$\mathbf{V_C} = \frac{V_0\angle 0^\circ}{(1 - \omega^2 LC + j\omega RC)} \tag{8.18}$$

Now we use the fact that $i_c = C dv_c/dt$ again. This means that

$$\mathbf{I} = j\omega C\mathbf{V_C}$$

Or using (8.18) in terms of the input voltage we have

$$\mathbf{I} = \frac{j\omega C}{(1 - \omega^2 LC + j\omega RC)} V_0\angle 0^\circ$$

Now we have a relation of the form $\mathbf{R} = \mathbf{HE}$, specifically where $\mathbf{H}$ is an admittance transfer function relating the excitation voltage to the response current **I.** So the transfer function is given by

$$\mathbf{H}(\omega) = \frac{j\omega C}{(1 - \omega^2 LC + j\omega RC)}$$

Now we do a little manipulation (multiply top and bottom by the complex conjugate of the denominator) to write this in terms of real and imaginary parts

$$\mathbf{H}(\omega) = \frac{j\omega C}{(1-\omega^2 LC + j\omega RC)} = \frac{j\omega C}{(1-\omega^2 LC + j\omega RC)} \frac{(1-\omega^2 LC - j\omega RC)}{(1-\omega^2 LC - j\omega RC)}$$

$$= \frac{\omega^2 C^2 R + j\omega C(1-\omega^2 LC)}{(1-\omega^2 LC)^2 + \omega^2 R^2 C^2}$$

Setting the imaginary part of this expression equal to zero, we find

$$\frac{\omega C(1-\omega^2 LC)}{(1-\omega^2 LC)^2 + \omega^2 R^2 C^2} = 0$$

$$\Rightarrow$$

$$1 - \omega^2 LC = 0$$

And we find that the resonant frequency is

$$\omega = \frac{1}{\sqrt{LC}}$$

EXAMPLE 8-3

Find the condition for resonance for the circuit shown in Fig. 8-6. The response of the circuit is the current flowing through the inductor L. The voltage source is $V_0 \cos \omega t$.

SOLUTION

We apply KVL to both panes in the circuit. On the left pane we have

$$V_0 \cos \omega t = R_A i_1 + R_B(i_1 - i_2) + v_c$$

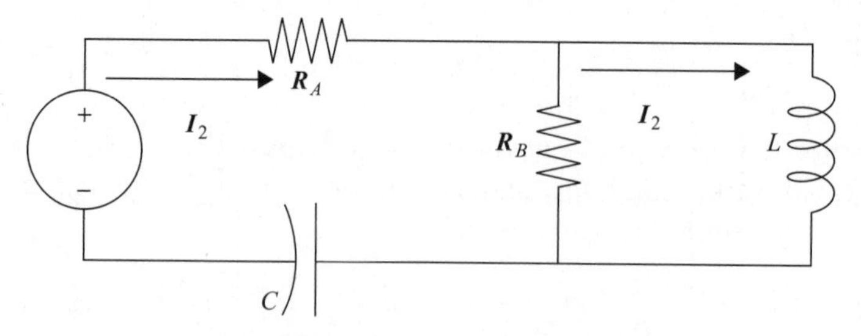

Fig. 8-6 Circuit for Example 8-3.

The current $i_1(t)$ is related to the voltage across the capacitor as

$$i_1(t) = C\frac{dv_c}{dt}$$

Or using phasors

$$\mathbf{I_1} = j\omega C\mathbf{V_c}$$

Hence

$$V_0\angle 0^\circ = \mathbf{I_1}\left(R_A + R_B + \frac{1}{j\omega C}\right) - R_B\mathbf{I_2}$$

Applying KVL to the pane on the right-hand side we obtain

$$R_B(i_2 - i_1) + v_L = 0$$

Using $v_L = L di_2/dt, \Rightarrow \mathbf{V_L} = j\omega L\mathbf{I_2}$ we obtain a second phasor equation

$$\mathbf{I_2} = \frac{R_B}{R_B + j\omega L}\mathbf{I_1}$$

Putting the two equations together and solving for $\mathbf{I_2}$ in terms of the source voltage we obtain

$$\mathbf{I_2} = \frac{\omega C R_B^2}{R_B}\frac{1}{\omega C R_B(R_A + R_B) + \omega L - 1 + J(\omega^2 LC(R_A + R_B) - R_B)}V_0\angle 0^\circ$$

The transfer function (an admittance) is

$$\mathbf{H}(\omega) = \frac{\omega C R_B^2}{R_B}\frac{1}{\omega C R_B(R_A + R_B) + \omega L - 1 + J(\omega^2 LC(R_A + R_B) - R_B)}$$

This can be written in the form $x + jy$ by multiplying by $\bar{z}/\bar{z}$, where z is the complex number in the denominator. That isn't necessary; we see that the phase angle of the transfer function will be zero if the coefficient of J vanishes, that is

$$(\omega^2 LC(R_A + R_B) - R_B) = 0$$

Solving for the frequency and taking the positive square root (frequencies are positive) we find

$$\omega = \sqrt{\frac{R_B}{LC(R_A + R_B)}}$$

This is the condition for resonance.

Filters

A *filter* is a circuit designed to allow certain frequencies to pass through to the output while blocking other frequencies. A filter is characterized by its transfer function $\mathbf{H}(\omega)$. There are four basic filter types, and we explore each of these in turn.

Suppose that you want a filter that only allows a response to an excitation if the frequency is greater than a critical frequency denoted ω_c. Since the filter allows high frequencies $\omega > \omega_c$ to pass through, we call this type of filter a *high-pass* filter. To illustrate the operation of the filter, we plot the magnitude of the transfer function $|\mathbf{H}(\omega)|$ against frequency. An example of an *ideal* high-pass filter is shown in Fig. 8-7. This is an ideal filter because the cutoff is sharp at the critical frequency; here we represent the transfer function by a unit step function. In a real filter, the cutoff is not sharp but is a rapidly (but smoothly) increasing curve that settles out at unity. Since the transfer function is zero if

$$\omega < \omega_c$$

Then the circuit does not respond to input for those frequencies at all.

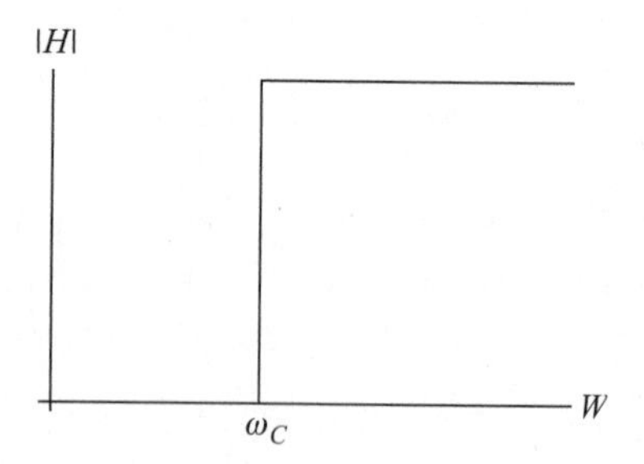

Fig. 8-7 A plot of the magnitude of the transfer function for an ideal high-pass filter.

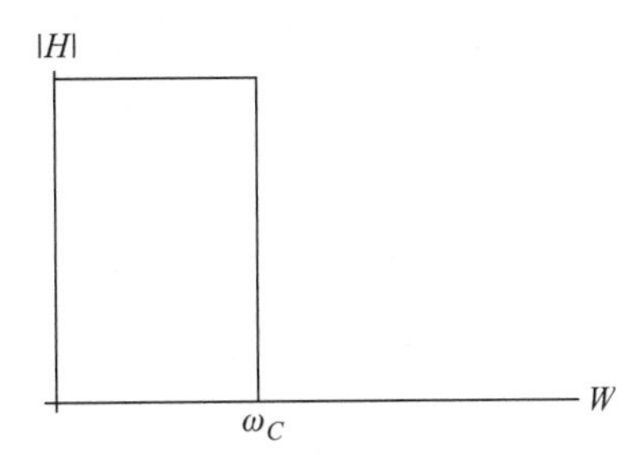

Fig. 8-8 An ideal low-pass filter.

A *low-pass filter* works in a manner opposite to a high-pass filter. This time, we only allow frequencies *below* the cutoff frequency to pass through. A plot of an ideal low-pass filter is shown in Fig. 8-8. If $\omega > \omega_c$, since the transfer function is zero there is no response to high-frequency input.

Next, we consider a *band-pass filter*. This type of filter will allow frequencies that fall within a certain frequency range or *band* to pass through, while blocking all others. The transfer function for a band-pass filter is shown in Fig. 8-9.

Finally, a *band-stop* filter allows most frequencies to pass through but blocks frequencies within a certain range. This is illustrated in Fig. 8-10.

EXAMPLE 8-4

Consider the circuit shown in Fig. 8-11 with the voltage source given by $v_s(t) = V_0 \cos \omega t$. The output of the circuit is the voltage across the resistor R. Show that this circuit functions as a high-pass filter and plot the magnitude of the transfer function.

SOLUTION

KVL in a clockwise direction around the loop in Fig. 8-11 gives

$$V_0 \cos \omega t = v_c + Ri$$

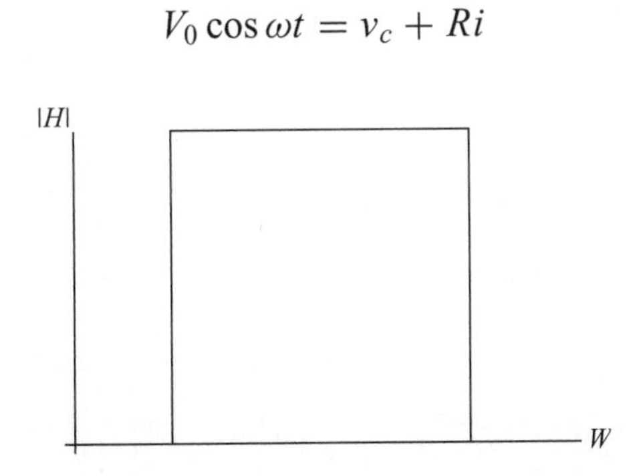

Fig. 8-9 The magnitude of the transfer function for an ideal band-pass filter. Frequencies that fall within a certain range are allowed to pass, while others are blocked.

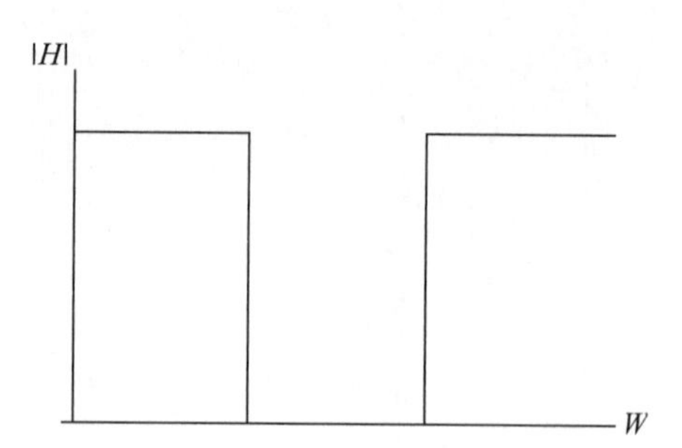

Fig. 8-10 A band-stop filter.

where v_c is the voltage across the capacitor and i is the current that flows through all the elements in the circuit since they are in series. By now we have memorized that

$$i_c = C\frac{dv_c}{dt}$$

So the current is described by the differential equation

$$V_0 \cos \omega t = v_c + RC\frac{dv_c}{dt}$$

Let's take the phasor transform of this equation to describe the problem in frequency space. We find

$$V_0\angle 0^\circ = \mathbf{V_c} + j\omega RC\mathbf{V_c}$$

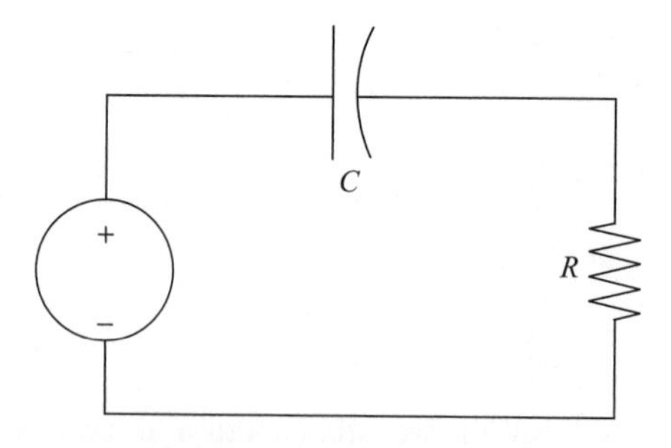

Fig. 8-11 A high-pass filter.

This immediately gives a solution for the voltage across the capacitor

$$\mathbf{V_c} = \frac{1}{1 + j\omega RC} V_0 \angle 0^\circ$$

Now the phasor current is related to the voltage across the capacitor in the following way

$$\mathbf{I} = j\omega C \mathbf{V_c} = \frac{j\omega C}{1 + j\omega RC} V_0 \angle 0^\circ$$

This is the same current flowing through the resistor since they are in series. So the output voltage across the resistor is, using Ohm's law

$$\mathbf{V}_R = \frac{j\omega RC}{1 + j\omega RC} V_0 \angle 0^\circ$$

Hence, the transfer function for the circuit is

$$\mathbf{H}(\omega) = \frac{j\omega RC}{1 + j\omega RC}$$

Multiplying and dividing by $1 - j\omega RC$ we obtain

$$\mathbf{H}(\omega) = \frac{\omega^2 R^2 C^2 + j\omega RC}{1 + \omega^2 R^2 C^2}$$

The magnitude of this expression is

$$|\mathbf{H}(\omega)| = \sqrt{\frac{\omega^4 R^4 C^4 + \omega^2 R^2 C^2}{(1 + \omega^2 R^2 C^2)^2}}$$

$$= \sqrt{\frac{\omega^2 R^2 C^2 (1 + \omega^2 R^2 C^2)}{(1 + \omega^2 R^2 C^2)^2}} = \frac{\omega RC}{\sqrt{1 + \omega^2 R^2 C^2}}$$

At this point we take a digression to learn how to estimate the cutoff frequency. This occurs when $|\mathbf{H}|$ is at half of its peak value, that is,

$$\frac{1}{\sqrt{2}} = \frac{1}{\sqrt{1 + \omega^2 R^2 C^2}}$$

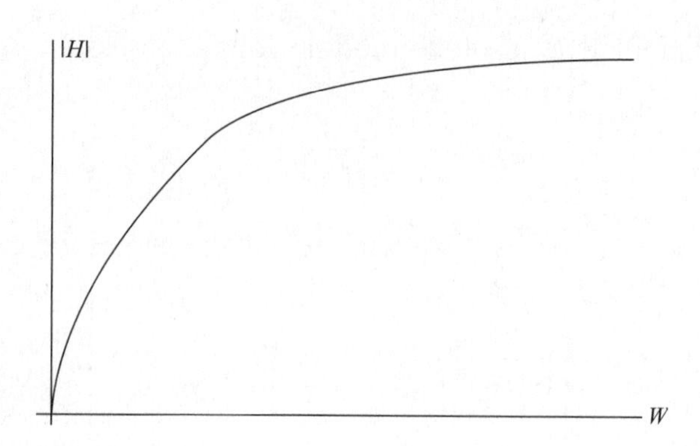

Fig. 8-12 The transfer function for the high-pass filter in Example 8-4.

This condition is known as *full-width at half power*. So the condition we need to solve is

$$2 = 1 + \omega_c^2 R^2 C^2$$

From which we find that

$$\omega_c = \frac{1}{RC}$$

So by tuning the values of the resistor and capacitor, we can construct a high-pass filter with the desired cutoff frequency. A plot of the transfer function is shown in Fig. 8-12.

Band-pass or band-stop filters can be constructed using RLC circuits. In a series RL circuit, the *damping parameter* is

$$\varsigma = \frac{R}{2L} \tag{8.19}$$

A filter is underdamped if

$$\varsigma < \omega_0 \tag{8.20}$$

It is critically damped if

$$\varsigma = \omega_0 \tag{8.21}$$

And it is overdamped if

$$\varsigma > \omega_0 \tag{8.22}$$

The bandwidth of the filter is

$$\Delta\omega = 2\varsigma = \frac{R}{L} \tag{8.23}$$

This gives the total frequency range about the critical frequency that is allowed to pass. That is, the filter will pass frequencies that lie within the range

$$\omega = \omega_c \pm \frac{\Delta\omega}{2} \tag{8.24}$$

The *quality* or Q-factor for the circuit is given by

$$Q = \frac{\omega_0}{\Delta\omega} = \frac{L}{R}\omega_0 \tag{8.25}$$

EXAMPLE 8-5
An RL filter has $R = 200\ \Omega$ and $L = 10\text{H}$. What is the bandwidth? If the resonant frequency is 100 Hz, what is the quality factor?

SOLUTION
Using (8.23) the bandwidth is

$$\Delta\omega = \frac{R}{L} = 20 \text{ rad/s}$$

The Q-factor is

$$Q = \frac{L}{R}\omega_0 = \left(\frac{10}{200}\right)(2\pi)(100) = 31.42$$

Summary

For a series LC circuit, the differential equation describing the voltage across the capacitor is

$$LC\frac{d^2v_C}{dt^2} + v_C = 0$$

The natural frequency of the circuit is given by

$$\omega_0 = \frac{1}{\sqrt{LC}}$$

A resonant frequency for an input is one where $\omega = \omega_0$. Resonant frequencies can cause the system to "blow up." The impedance of a circuit is given by

$$\mathbf{Z} = \sqrt{R^2 + X^2} \angle \tan^{-1}\left(\frac{X}{R}\right)$$

We can use this to write "Ohm's law" as $\mathbf{V} = \mathbf{ZI}$. The susceptance is

$$\mathbf{Y} = \frac{1}{\mathbf{Z}}$$

Given an excitation $\mathbf{E}$ we can describe the behavior of a circuit in terms of the transfer function $\mathbf{H}$

$$\mathbf{R} = \mathbf{HE}$$

where $\mathbf{R}$ is the response of the circuit.

Quiz

1. A load has a voltage $\mathbf{V} = 20\angle 0°$ and current $\mathbf{I} = 2\angle 20°$. Find the impedance and determine a series circuit that will model the load. Is the circuit inductive or capacitive? Assume that $\omega = 100$ rad/s.

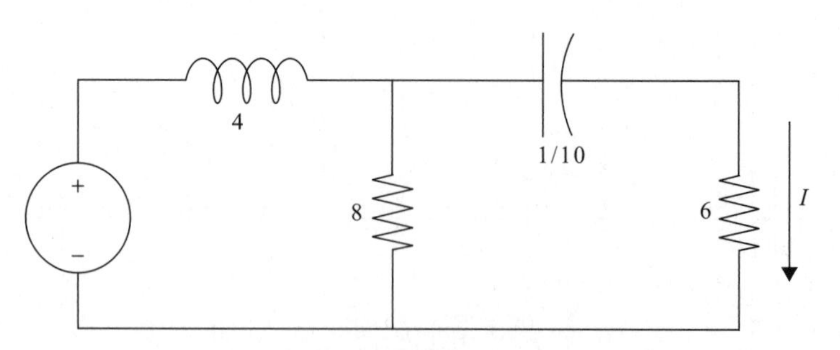

Fig. 8-13 Circuit diagram for Problem 2.

2. Consider the circuit shown in Fig. 8-13. The response of the circuit is the current flowing through the 6 Ω resistor. Determine the resonant frequency if $v_s(t) = 10 \cos \omega t$.
3. Consider the circuit in Fig. 8-6. Suppose that the voltage source is replaced by a current source $i_s(t) = I_0 \cos \omega t$ and the positions of the inductor and capacitor are switched. What is the resonant frequency?
4. Reverse the positions of the capacitor and resistor in Fig. 8-11. Does the circuit still function as a filter?
5. An RLC filter has $R = 100\ \Omega$ and $L = 4$ H. What is the bandwidth? If the resonant frequency is 100 Hz, what is the quality factor?

CHAPTER 9

Operational Amplifiers

An *operational amplifier* or *op amp* is a circuit that takes an input voltage and *amplifies* it. The symbol used to represent an op amp in a circuit diagram is shown in Fig. 9-1.

An op amp is defined by two simple equations. The first thing to note is that the voltage across the input terminals is zero. Hence

$$V_a = V_b \tag{9.1}$$

The second relation that is essential for analyzing op amp circuits is that the currents drawn at a and b in Fig. 9-1 are zero

$$I_a = I_b = 0 \tag{9.2}$$

Despite this, we will see that the op amp will result in voltage gains at the output terminal c. How does this work? Two voltages are input to terminals a and b. Their difference is then amplified and output at c, which is taken with referenc to ground. Although we won't worry about the internal construction of an op amp, note that it consists of a set of resistors and dependent voltage

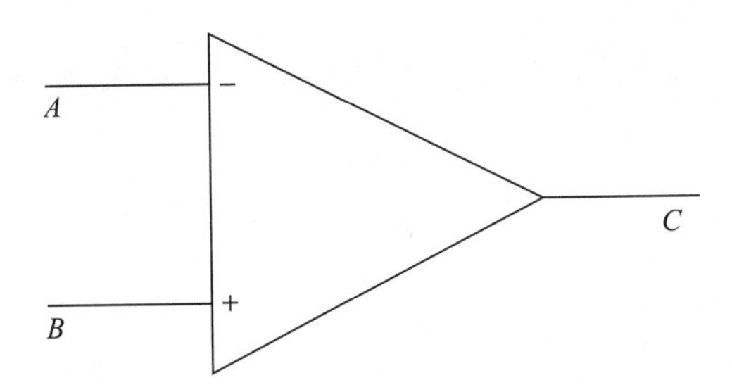

Fig. 9-1 An operational amplifier.

source. The internal voltage source is related to the input voltages by

$$A(V_+ - V_-) \tag{9.3}$$

The constant A is known as the *open-loop voltage gain.* To see how op amp circuits work, it's best to examine some popular example circuits. When analyzing op amp circuits, remember to take the input voltage across the op amp terminals to be zero and that the op amp draws zero current. The analysis is then reduced to applying KVL and KCL to the circuit elements connected to the op amp.

The Noninverting Amplifier

In Fig. 9-2, we show a circuit that is called a *noninverting amplifier.* This circuit will take the input voltage V_{in} and amplify it at the output V_{out}. This is illustrated in Fig. 9-2.

The output voltage of this circuit is the voltage across the load resistor R_L. To calculate it we note that

- The voltage across the input terminals of the op amp is essentially zero.
- The voltage across the input resistor R_i is the input voltage V_{in}.

Applying KVL about the resistors gives

$$V_f = V_{\text{in}} - V_{\text{out}}$$

Now apply KCL to the node where R_f and R_i meet. We have

$$I_i + I_f = 0$$

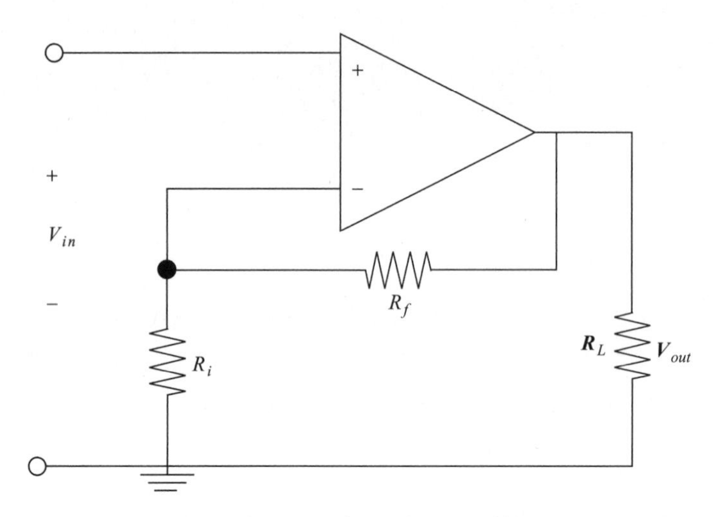

Fig. 9-2 A noninverting amplifier.

Using Ohm's law

$$I_f = \frac{V_f}{R_f} = \frac{V_{\text{in}} - V_{\text{out}}}{R_f}$$

And

$$I_{\text{in}} = \frac{V_{\text{in}}}{R_i}$$

Therefore $I_i + I_f = 0$ gives us

$$\frac{V_{\text{in}}}{R_i} = \frac{V_{\text{out}} - V_{\text{in}}}{R_f}$$

Hence

$$V_{\text{out}} = V_{\text{in}} \left(1 + \frac{R_f}{R_i}\right) \tag{9.4}$$

We call the factor

$$1 + \frac{R_f}{R_i} \tag{9.5}$$

the *closed-loop gain* of the amplifier.

EXAMPLE 9-1

Consider a noninverting amplifier with an input voltage of 10 V, $R_f = 400\ \Omega$, $R_i = 20\ \Omega$, and $R_L = 10\ \Omega$. Determine the closed-loop gain and the output voltage.

SOLUTION

Using (9.5) we find that the closed-loop gain is

$$1 + \frac{R_f}{R_i} = 1 + \frac{400}{20} = 21$$

This is a dimensionless quantity. The output voltage can be found by using (9.4). In this case we have

$$V_{\text{out}} = V_{\text{in}} \left(1 + \frac{R_f}{R_i}\right) = (10\ \text{V})(21) = 210\ \text{V}$$

Inverting Amplifier

An *inverting amplifier* reverses the sign of the input voltage. A circuit that will implement an inverting amplifier is shown in Fig. 9-3.

Again, the voltage across the input terminals of the op amp is zero. Therefore, the voltage across the resistor R_i is by KVL V_i. The output voltage V_{out} actually is across the resistor R_f. Applying KCL at the node connecting R_i and R_f gives

$$\frac{V_i}{R_i} + \frac{V_0}{R_f} = 0$$

Therefore the output voltage for an inverting amplifier as shown in Fig. 9-3 is

$$V_{\text{out}} = -\frac{R_f}{R_i} V_{\text{in}} \tag{9.6}$$

EXAMPLE 9-2

Consider an inverting amplifier with $R_f = 200\ \Omega$ and $R_i = 10\ \Omega$. If the input voltage supplied to the amplifier is 10 V, what is the output voltage?

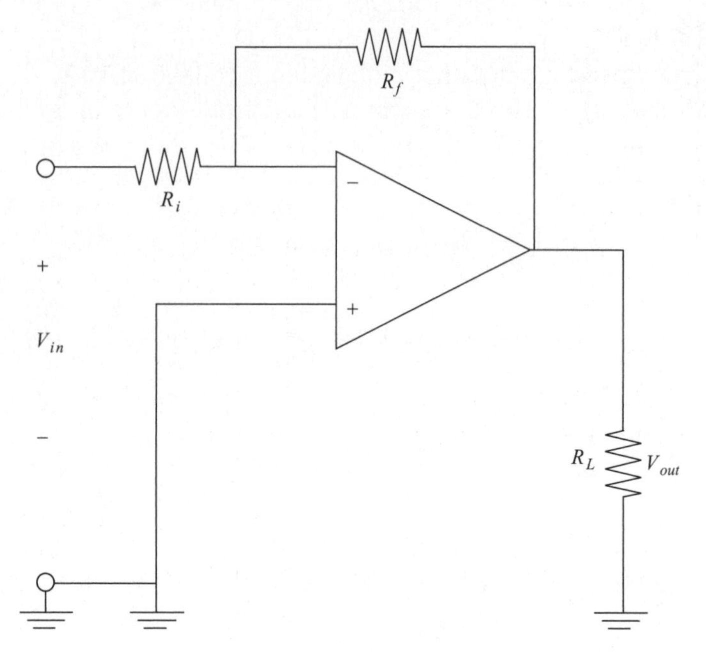

Fig. 9-3 An inverting amplifier.

SOLUTION
Using (9.6) we find the output voltage is

$$V_{\text{out}} = -\frac{R_f}{R_i} V_{\text{in}} = -\frac{200}{10}(20 \text{ V}) = -400 \text{ V}$$

The Summing Amplifier

The final example of an op amp circuit we consider is called a summing amplifier or summer. This type of amplifier sums up multiple voltages to produce an output voltage. Specifically, it can be shown that the circuit shown in Fig. 9-4 produces the output voltage

$$V_{\text{out}} = -R_f \left(\frac{V_a}{R_a} + \frac{V_b}{R_b} + \frac{V_c}{R_c} \right) \tag{9.7}$$

Hence, a summer amplifies each voltage, adds them up, and inverts the output. In the special case where $R_a = R_b = R_c = R_f$, then we get a circuit that directly

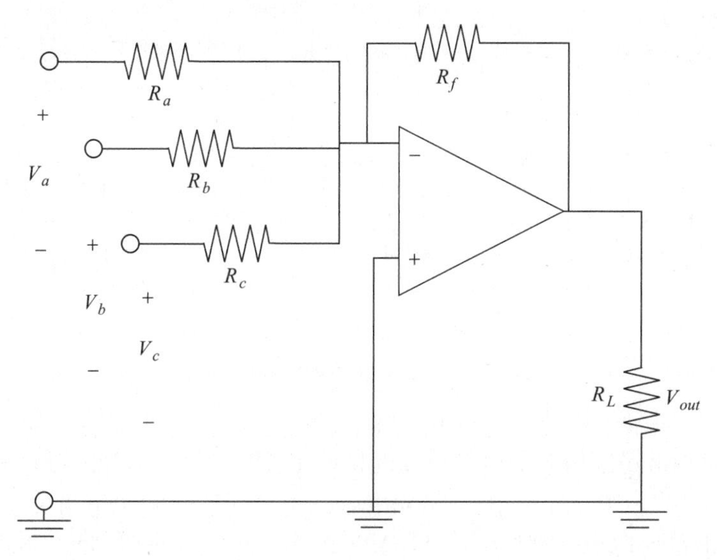

Fig. 9-4 A summing amplifier scales, adds up, and then inverts the sum of the input voltages.

adds up the voltages

$$V_{\text{out}} = -(V_a + V_b + V_c) \tag{9.8}$$

EXAMPLE 9-3

A summing amplifier has $R_a = 200\ \Omega$. Find the remaining resistances such that $V_{\text{out}} = -(12V_a + V_b + 4V_c)$.

SOLUTION

We can solve this problem by using (9.7). First we find R_f

$$12 = \frac{R_f}{R_a}, \Rightarrow R_f = (12)(200\ \Omega) = 2400\ \Omega$$

Since the coefficient of V_b is 1, then $R_b = R_f = 2400\ \Omega$. For the remaining term, we find

$$4 = \frac{R_f}{R_c}, \Rightarrow R_c = \frac{2400\ \Omega}{4} = 600\ \Omega$$

Summary

An operational amplifier has two input terminals, and the voltage at each terminal is the same. In addition, both terminals draw zero current. A noninverting amplifier takes the voltage at the input terminals and steps it up to an amplified voltage at the output terminal. An inverting amplifier increases the magnitude of the voltage but changes the sign.

Quiz

1. Consider an inverting amplifier with $R_f = 1000\ \Omega$ and $R_i = 50\ \Omega$. If the input voltage supplied to the amplifier is 10 V, what is the output voltage?
2. Consider an noninverting amplifier with $R_f = 1000\ \Omega$ and $R_i = 50\ \Omega$. If the input voltage supplied to the amplifier is 10 V. What is the closed-loop gain?
3. For the noninverting amplifier in Problem 2, what is the output voltage?
4. Three voltages are input to a summer as $V_a = 2$ V, $V_b = -3$ V, and $V_c = 8$ V. What is the output voltage?

CHAPTER 10

Sinusoidal Steady-State Power Calculations

In this chapter we study power in circuits in more detail. In particular, our focus will be power in circuits with sinusoidal sources. We begin by considering maximum power transfer.

Maximum Power Transfer

Let's look at an arbitrary circuit consisting of sources and resistors that is set up to deliver power to some load. This is illustrated schematically in Fig. 10-1.

We consider the special case where the load network is a single resistor called the *load resistor*. To determine the power delivered by the source network, we can always use Thevenin's theorem to replace it by a simple Thevenin equivalent circuit at the terminals A–B, using the techniques we learned in Chapter 3. Therefore, with a single load resistor and using Thevenin's theorem, we can

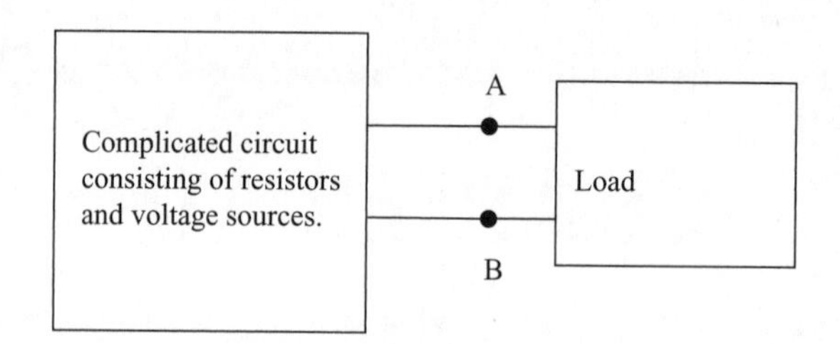

Fig. 10-1 An arbitrary circuit consisting of sources and resistors used to deliver power to a load.

replace the arbitrary network shown in Fig. 10-1 with the simple network shown in Fig. 10-2.

Now let's determine the power absorbed by the load. It's easy to show that, given the circuit shown in Fig. 10-2, the power delivered to the load is

$$P_L = I^2 R_L = \left(\frac{V_{\text{TH}}}{R_{\text{TH}} + R_L} \right)^2 R_L \tag{10.1}$$

Now we want to find out what the value of the load resistance R_L is that will maximize the power delivered to the load. We can use (10.1) and some basic calculus to find out. The maximum power transfer occurs for the value of R_L that satisfies

$$\frac{dP}{dR_L} = 0 \tag{10.2}$$

Let's generate a rough plot of (10.1). In Fig. 10-3, we see that the power will vary with load resistance in a way that looks roughly like a skewed Bell curve.

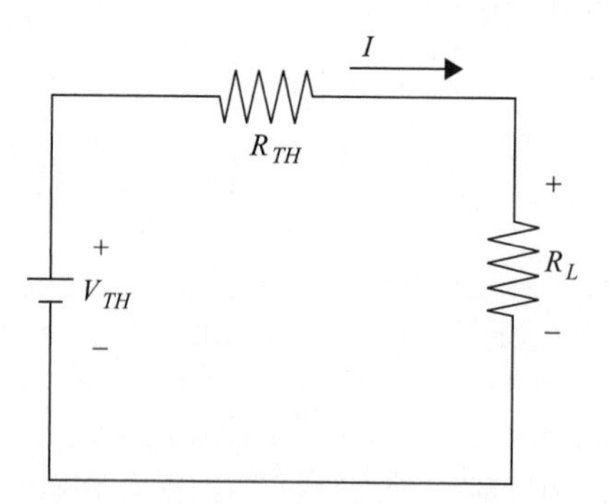

Fig. 10-2 The network in Fig. 10-1 replaced by its Thevenin equivalent with a load resistor.

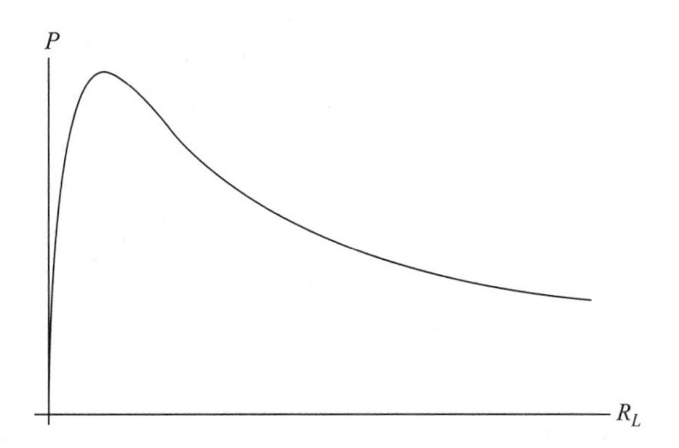

Fig. 10-3 A plot of power delivered with load resistance.

To calculate (10.2), recall that

$$\left(\frac{f}{g}\right)' = \frac{f'g - g'f}{g^2}$$

We take

$$f = R_L, \quad g = (R_{\text{TH}} + R_L)^2$$

Then

$$f' = 1, \quad g' = 2(R_{\text{TH}} + R_L)$$

So we have

$$\frac{dP}{dR_L} = V_{\text{TH}}^2 \left[\frac{(R_{\text{TH}} + R_L)^2 - 2R_L(R_{\text{TH}} + R_L)}{(R_{\text{TH}} + R_L)^4}\right] = 0$$

We cancel the term

$$\frac{V_{\text{TH}}^2}{(R_{\text{TH}} + R_L)^4}$$

Giving

$$(R_{\text{TH}} + R_L)^2 - 2R_L(R_{\text{TH}} + R_L) = 0$$

Solving, we find that the load resistance that results in maximum power transfer is

$$R_L = R_{TH} \tag{10.3}$$

Therefore to maximize power transfer we set the load resistance equal to the Thevenin resistance. Using (10.1), we see that the power transferred is

$$P_L = \frac{V_{TH}^2}{4R_{TH}} \tag{10.4}$$

EXAMPLE 10-1

A load resistor is connected to the circuit shown in Fig. 10-4. What value of load resistance should be used to maximize the power transfer if $R_1 = 2\ \Omega$, $R_2 = 3\ \Omega$, $R_3 = R_4 = 6\ \Omega$, the voltage source is $V_s = 15$ V, and the current is $I_L = 3$ A? What power is transferred?

SOLUTION

We already calculated the Thevenin equivalent circuit in Example 3-5. This is shown in Fig. 10-5, where we see that $V_{TH} = 1.25$ V and $R_{TH} = 4.2\ \Omega$.

The load resistance that maximizes the power is R_{TH}

$$R_L = 4.2\ \Omega$$

The power delivered is

$$P_L = \frac{V_{TH}^2}{4R_{TH}} = \frac{(1.25)^2}{4(4.2)} = 0.09\ \text{W}$$

Perhaps this isn't a very useful circuit!

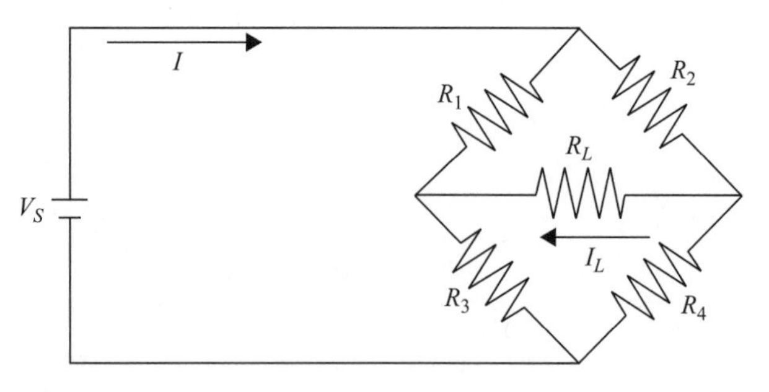

Fig. 10-4 In Example 10-1 we find the value of R_L that will maximize the power transfer in this circuit.

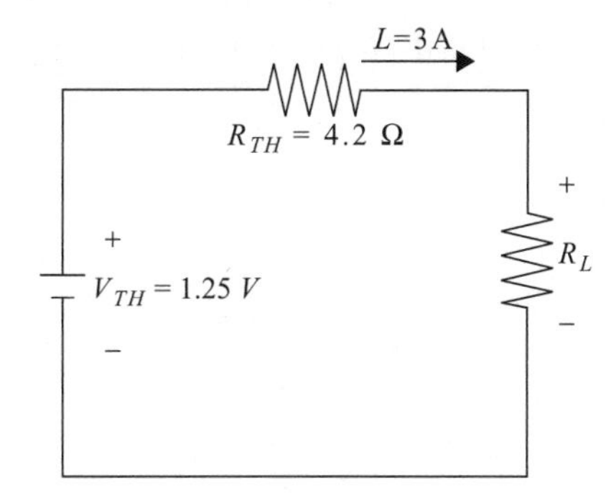

Fig. 10-5 The Thevenin equivalent circuit for the network shown in Fig. 10-4, derived in Example 3-5.

Instantaneous Power

In most cases the load is more complicated than a simple resistor. The load will not contain sources, but it will consist of resistors, capacitors, and inductors. A current will flow from the power source to the load through a two-terminal connection. This is illustrated in Fig. 10-6.

We can characterize the load by its impedance, which can be written in terms of a resistance R and reactance X as

$$Z = R + jX \tag{10.5}$$

In polar form, $Z = |Z|\, e^{j\phi} = |Z| \angle\phi$. We can write the voltage in terms of its root mean square or effective value across the terminals A-B due to the power source as

$$v(t) = V_0 \cos(\omega t + \psi) \tag{10.6}$$

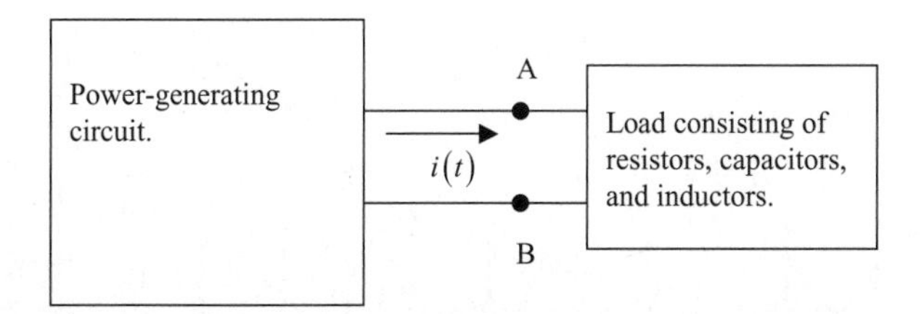

Fig. 10-6 A schematic representation of a general power load consisting of passive circuit elements.

The phasor transform of the input voltage is

$$\mathbf{V} = V_0\angle\psi \tag{10.7}$$

We can calculate the current $i(t)$ delivered to the load by using Ohm's law. If we denote the phasor transform of the current as **I** then we have

$$\mathbf{V} = \mathbf{ZI}, \Rightarrow \mathbf{I} = \frac{V_0\angle\psi}{|Z|\angle\phi} = \frac{V_0}{|Z|}\angle(\psi - \phi) \tag{10.8}$$

The amplitude of the current is

$$I_0 = \frac{V_0}{|Z|}$$

Given (10.6) and (10.8), we deduce that the current in the time domain has the form

$$i(t) = I_0 \cos(\omega t + (\psi - \phi)) \tag{10.9}$$

If ϕ is positive, then we see that the current delivered to the load lags the voltage. The *instantaneous power* delivered to the load is

$$P(t) = v(t)i(t) \tag{10.10}$$

Using (10.6) and (10.9) this becomes

$$P(t) = V_0 \cos(\omega t + \psi) I_0 \cos(\omega t + (\psi - \phi)) \tag{10.11}$$

To simplify this expression, we can use the trig identities

$$\cos A \cos B = \frac{1}{2}\left[\cos(A - B) + \cos(A + B)\right]$$

Letting $A = \omega t + \psi, \ B = \omega t + \psi - \phi$ in (10.11), we have

$$A - B = \phi,$$

$$A + B = 2\omega t + 2\psi - \phi$$

Hence, the instantaneous power delivered to the load can be written as

$$P(t) = \frac{V_0 I_0}{2} \cos(\phi) + \frac{V_0 I_0}{2} \cos(2\omega t - \phi) \tag{10.12}$$

Average and Reactive Power

If we calculate the average of (10.12), the second term will wash out. Therefore it's easy to see that the *average power* delivered to the load is given by the first term, that is

$$P_{\text{av}} = \frac{V_0 I_0}{2} \cos(\phi) \tag{10.13}$$

The *power factor* is given by

$$\text{p.f.} = \cos\phi \tag{10.14}$$

Notice that $0 \leq \cos\phi \leq 1$. This tells us that the power factor is a measure of the efficiency at which power is delivered to the load. In addition, note that it includes the phase angle ϕ, which is the phase difference between the voltage and current. The extreme values of the power factor tell us

- If the power factor is 1, then the voltage and current are in phase.
- If the power factor is 0, then the voltage leads or lags the current by 90°. In that case, no average power is delivered to the load.

EXAMPLE 10-2

A power source with $v(t) = 100\cos 50t$ is connected to a load consisting of a resistor $R = 20$ and inductor $L = 4$ connected in series, as shown in Fig. 10-7. Find the current flowing through the load, the instantaneous power, the average power, and the power factor for this circuit.

SOLUTION

The impedance of the load is

$$Z = R + jX$$

where $X = \omega L$ is the reactance of the inductor. Using the values provided in the problem statement we have

$$Z = 20 + j(50)(4) = 20 + j200$$

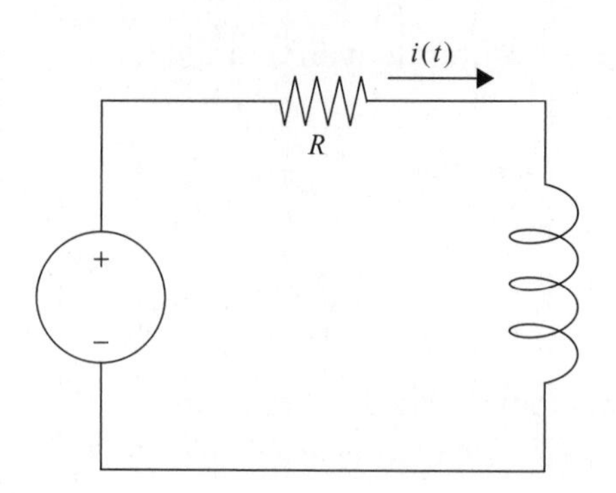

Fig. 10-7 A load consisting of a resistor and inductor connected in series is connected to a sinusoidal power source.

The phase angle of the load impedance is

$$\phi = \tan^{-1}\left(\frac{\omega L}{R}\right) = \tan^{-1}\left(\frac{200}{20}\right) = 84^{\circ}$$

The magnitude of the impedance is

$$|Z| = \sqrt{R^2 + (\omega L)^2} = \sqrt{(20)^2 + (200)^2} = 201$$

Hence, the polar representation of the load impedance is

$$Z = 201\angle 84^{\circ}$$

In frequency space, the current flowing through the load is

$$I = \frac{V}{Z} = \frac{400\angle 0^{\circ}}{201\angle 84^{\circ}} = 2\angle{-84^{\circ}}$$

As a function of time the current flowing through the load is

$$i(t) = 2\cos(50t - 84^{\circ})$$

The instantaneous power using (10.12) is

$$\begin{aligned} P(t) = v(t)i(t) &= \frac{(400)(2)}{2}\cos(84^{\circ}) + \frac{(400)(2)}{2}\cos(100t + 84^{\circ}) \\ &= 42 + 400\cos(100t + 84^{\circ})\ \text{W} \end{aligned}$$

The average power using (10.13) is

$$P_{av} = \frac{(V_0)(I_0)}{2} \cos\phi = \frac{(400)(2)}{2} \cos(84^\circ) = 42 \text{ W}$$

Finally, the power factor for this circuit is

$$\text{p.f.} = \cos\phi = \cos(84^\circ) \approx 0.10$$

The RMS Value and Power Calculations

The power delivered to the load is often characterized in terms of the RMS or effective values. To quickly review, if $v(t) = V_0 \cos(\omega t + \phi)$ and $i(t) = I_0 \cos(\omega t + \theta)$, then the effective voltage and current are

$$V_{eff} = \frac{V_0}{\sqrt{2}}, \quad I_{eff} = \frac{I_0}{\sqrt{2}} \tag{10.15}$$

In terms of the effective values, the average power is

$$P_{av} = \frac{V_0 I_0}{2} \cos\phi = \frac{(\sqrt{2}V_{eff})(\sqrt{2}I_{eff})}{2} \cos\phi = V_{eff} I_{eff} \cos\phi \tag{10.16}$$

where $\cos\phi$ is the power factor. Let's quickly review how to find the effective value for the voltage of a household outlet and then see how to use average power in practice.

EXAMPLE 10-3
The voltage of an ordinary outlet is $v(t) = 120 \sin 377t$. What is the effective voltage?

SOLUTION
The effective voltage is

$$V_{eff} = \frac{V_0}{\sqrt{2}} = \frac{170}{\sqrt{2}} = 120 \text{ V}$$

EXAMPLE 10-4
An electrical device is rated at 120 V, 220 W, with a power factor of 0.7 lagging. Describe the makeup of the device in terms of passive circuit elements.

SOLUTION

Assuming the device is connected to an ordinary outlet, we have the phasor representation of the voltage given by

$$\mathbf{V} = 170\angle 0^\circ$$

Rewriting this in terms of the effective voltage, this is

$$\mathbf{V} = 120(\sqrt{2})\angle 0^\circ$$

The power given is the average power. We can use this to find the effective value of the current using (10.16). That is,

$$P_{\text{av}} = V_{\text{eff}} I_{\text{eff}} \cos\phi$$

Therefore

$$I_{\text{eff}} = \frac{P_{\text{av}}}{V_{\text{eff}}\cos\phi} = \frac{220}{(120)(0.7)} = 2.6 \text{ A}$$

The amplitude of the current is $I_0 = \sqrt{2}\, I_{\text{eff}} = \sqrt{2}(2.6) = 3.7$ A. Next, we need to find the phase angle of the current. This is done by inverting the power factor. We add a negative sign because we are told that the device is lagging

$$\phi = -\cos^{-1}(0.7) = -46^\circ$$

Hence, the polar representation of the current is

$$\mathbf{I} = I_0\angle -46^\circ = 3.7\angle -46^\circ$$

The impedance of the circuit representing the device is

$$\mathbf{Z} = \frac{\mathbf{V}}{\mathbf{I}} = \frac{170\angle 0^\circ}{3.7\angle -46^\circ} = 46\angle 46^\circ = 46e^{j46^\circ}$$

Now we need to convert this into a Cartesian representation to determine the resistance and admittance. We can use Euler's identity

$$e^{j\theta} = \cos\theta + j\sin\theta$$

Since the angle is so close to 45°, we can use this value without adding much error and so take

$$\cos\theta = \sin\theta = \frac{1}{\sqrt{2}}$$

Hence

$$Z = 46\left(\frac{1}{\sqrt{2}} + j\frac{1}{\sqrt{2}}\right) = 32.5 + j32.5 = R + jX$$

We see immediately that $R = 32.5$. For the admittance, we can model the circuit with an inductor so take

$$X = \omega L = 377L$$
$$\Rightarrow L = \frac{X}{377} = \frac{32.5}{377} = 86 \text{ mH}$$

Our model, then, of the device is a 32.5 Ω resistor connected in series with a 86 mH inductor.

EXAMPLE 10-5

A load is connected in parallel across a $V_{\text{eff}} = 300$ V power source. The load is rated at 200 W with a power factor given as p.f. = 0.8 lagging. Find the effective current flowing through the load and describe a circuit that can model the load. Assume that $\omega = 100$.

SOLUTION

The circuit is shown in Fig. 10-8.

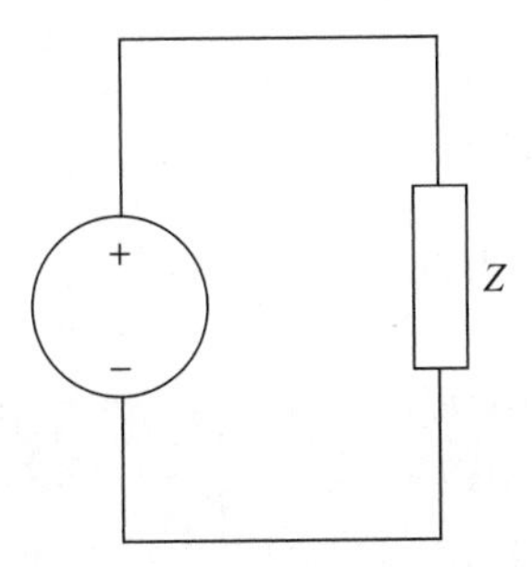

Fig. 10-8 The circuit studied in Example 10-4.

We can find the current by using the relation for average power (10.16). This time we have

$$I_{\text{eff}} = \frac{P_{\text{av}}}{V_{\text{eff}} \cos \phi} = \frac{200}{(300)(0.8)} = 0.83 \text{ A}$$

Again the load is lagging so the angle is

$$\phi = -\cos^{-1}(0.8) = -37^\circ$$

The impedance is

$$\mathbf{Z} = \frac{\mathbf{V}}{\mathbf{I}} = \frac{300\angle 0^\circ}{0.83(\sqrt{2})\angle -37^\circ} = 255\angle 37^\circ = 255 e^{j37^\circ}$$

Hence

$$Z = 203.65 + j153.46$$

We can model the load as a resistor with $R = 203.65$ in series with an inductor. Taking $\omega = 100$ we have

$$\omega L = 153, \Rightarrow$$
$$L = \frac{153}{100} = 1.53 \text{ H}$$

Next we consider an example with a capacitive load.

EXAMPLE 10-6

A current $i(t) = 40 \cos(100t + 20^\circ)$ is delivered to a capacitor $C = 1/10$. Find the instantaneous and average power.

SOLUTION

The phasor representation of the current is

$$\mathbf{I} = I_0 \angle \phi_i = 40\angle 20^\circ$$

For a capacitor we have the relation in frequency space

$$\mathbf{V} = \frac{1}{j\omega C}\mathbf{I}$$

With a purely capacitive load, the voltage and current are 90° out of phase. The relationship is

$$\phi_i = \phi_v + 90^\circ$$

In this case

$$20^\circ = \phi_v + 90^\circ, \Rightarrow \phi_v = -70^\circ$$

Hence

$$\mathbf{V} = \frac{1}{j(100)(1/10)} 40\angle{-70^\circ} = -j4\angle{-70^\circ}$$

Now, notice that

$$-j4e^{j70^\circ} = -j4(\cos(-70^\circ) + j\sin(-70^\circ)) = -j4\cos(-70^\circ) + 4\sin(-70^\circ)$$

To write down the voltage in the time domain, we take the real part of this expression. Therefore, in the time domain the voltage is

$$v(t) = 4\sin(100t - 70^\circ)$$

The voltage and current are 90° out of phase. This is shown in Fig. 10-9.

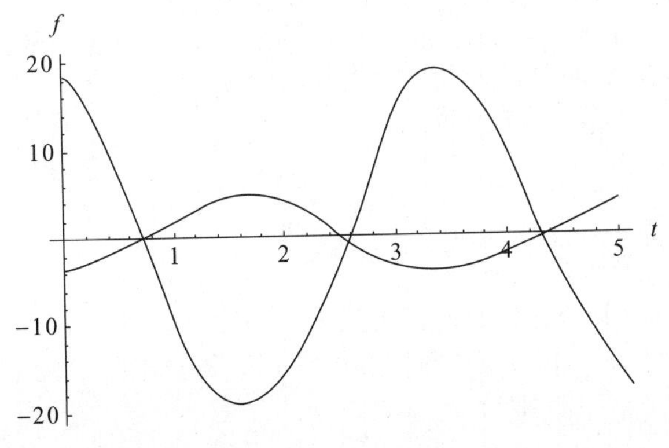

Fig. 10-9 The voltage (the curve with the smaller amplitude) and current in Example 10-5 are 90° out of phase. The current is not drawn to scale.

The instantaneous power is

$$P(t) = v(t)i(t) = \frac{(4)(40)}{2}\cos(\phi_v - \phi_i) + \frac{(4)(40)}{2}\cos(200t + \phi_v + \phi_i)$$
$$= 80\cos(200t - 50^\circ)$$

The average power is

$$P_{\text{av}} = \frac{(4)(40)}{2}\cos(90^\circ) = 0$$

Here we see an important result—a purely capacitive load is *lossless* when the power delivered is sinusoidal. This means that there is zero average power. Looking at the instantaneous power, we see that it's a pure sinusoid, meaning that power goes positive and negative. In other words, power flows back and forth from the source to the capacitor. This is why we refer to a capacitor as a *reactive* component.

A rule of thumb to remember is that capacitors create or generate reactive power. In contrast, inductors absorb reactive power. Loads tend to be inductive, so a corrective capacitor is often included in parallel with the power source to maximize the power delivered to the load. To see this, consider the circuit shown in Fig. 10-7 and studied in Example 10-1, in which we found the power factor to be a measly 0.1. This time we will insert a capacitor in parallel with the voltage source before delivering power to the load. A capacitor used in this way is called a *power corrective* or *shunt* capacitor. What value should be used for the capacitance to maximize the power factor?

First, we need to compute the impedance of the circuit including the capacitor. Remember, for a capacitor we have

$$X = \frac{1}{j\omega C}$$

Impedances work just like resistances when considering elements in series and parallel. The original impedance due to the series resistor-inductor combination is

$$Z_1 = R + j\omega L$$

The impedance due to the capacitor is

$$Z_2 = \frac{1}{j\omega C}$$

If these two impedances are in parallel, then the total impedance is

$$\frac{1}{Z} = \frac{1}{Z_1} + \frac{1}{Z_2} = \frac{1}{R + j\omega L} + j\omega C = \frac{1}{R + j\omega L} + \frac{j\omega C(R + j\omega L)}{R + j\omega L}$$

$$= \frac{1 - \omega^2 LC + j\omega RC}{R + j\omega L}$$

Hence

$$Z = \frac{R + j\omega L}{1 - \omega^2 LC + j\omega RC}$$

The ideal case would be to have $\phi = 0$ leading to a p.f. $= 1$. This will be true when the phase angle for the expression in the numerator matches the phase angle for the expression in the denominator, that is,

$$\tan^{-1}\left(\frac{\omega L}{R}\right) = \tan^{-1}\left(\frac{\omega RC}{1 - \omega^2 LC}\right)$$

So we must have

$$\frac{\omega L}{R} = \frac{\omega RC}{1 - \omega^2 LC}$$

Solving for C we find

$$C = \frac{L}{\omega^2 L^2 + R^2}$$

For the values used in Example 10-1, $R = 20$, $L = 4$, $\omega = 50$, the capacitance should be 99 μF.

Complex Power

When examining the power delivered to a load we can break the power down into three distinct quantities. We have already seen one of these, the average power delivered to the load

$$P = \frac{V_0 I_0}{2} \cos\phi = V_{\text{eff}} I_{\text{eff}} \cos\phi \tag{10.17}$$

We have denoted it by P in this context. Sometimes this is known as the *real power.* It is related to two other measures of power known as the *reactive power* and the *complex power.* This is done using a device known as an *impedance triangle* that relates the various components of the impedance to its magnitude. Let's start by writing down the expression for impedance

$$Z = R + jX \tag{10.18}$$

We can think of this as a "vector" in the complex plane. The x component of the vector is just R and the y component of this vector is X. If we draw R as a vector lying entirely along the x axis from the origin and jX as a vector lying parallel to the y axis with its tail at the head of R, we can connect the two "vectors" by Z, which is found by vector addition. This forms a triangle called an impedance triangle, which we show in Fig. 10-10.

Now we consider this triangle by letting each term go to a power expression, which can be done by multiplying each term by I^2. This results in a *power triangle,* which is shown in Fig. 10-11.

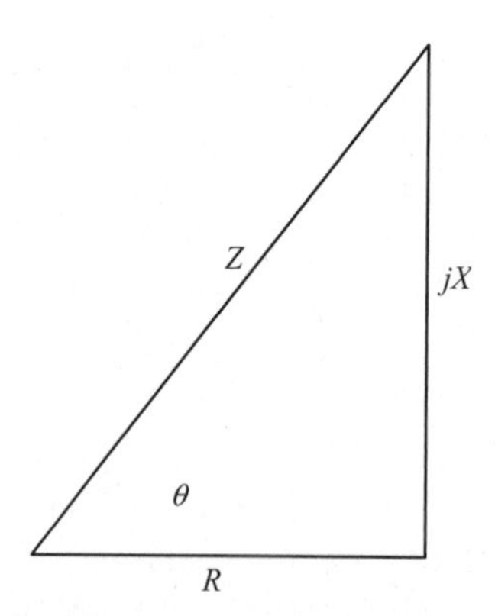

Fig. 10-10 An impedance triangle.

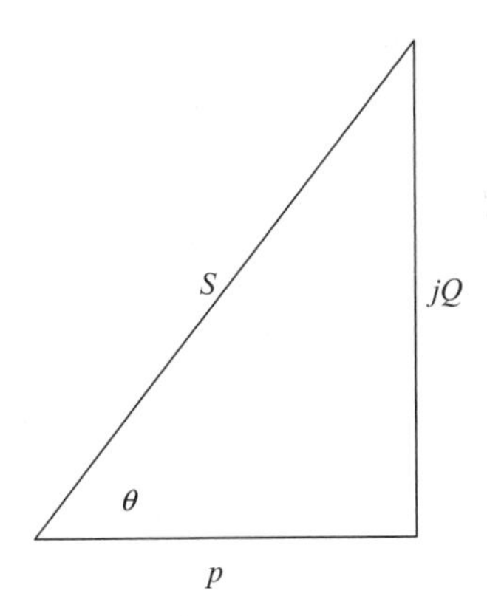

Fig. 10-11 A power triangle.

Each of the terms is defined as follows

$$S = I^2 Z, \quad P = I^2 R, \quad Q = I^2 X \tag{10.19}$$

Hence

$$S = P + jQ \tag{10.20}$$

As noted above, P is just the average power that we have used so far. The *reactive power* is given by Q. Using the angle ϕ by which the input voltage leads the input current the reactive power is defined as

$$Q = V_{\text{eff}} I_{\text{eff}} \sin \phi \tag{10.21}$$

The last component of the power triangle is given by S, which is called the *apparent power.* It is given this name because it is simply the product of the effective voltage and current, $|S| = V_{\text{eff}} I_{\text{eff}}$. This is easy to derive by applying the Pythagorean theorem to the triangle in Fig. 10-11. The magnitude of the apparent power is

$$|S| = \sqrt{P^2 + Q^2} = \sqrt{(V_{\text{eff}} I_{\text{eff}} \cos \phi)^2 + (V_{\text{eff}} I_{\text{eff}} \sin \phi)^2} = V_{\text{eff}} I_{\text{eff}}$$

Summary

To maximize power transfer we set the load resistance equal to the Thevenin equivalent resistance. If a circuit consists of more complicated elements, such as

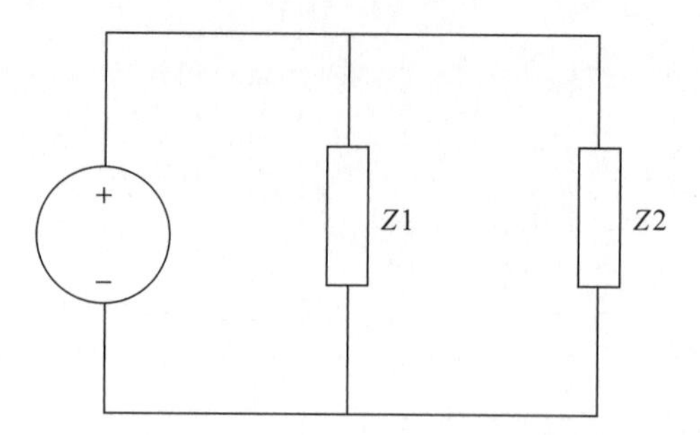

Fig. 10-12 A power source with two loads connected in parallel.

capacitors and resistors, we need to calculate the instantaneous or average power. The instantaneous power delivered to a load is $P(t) = v(t)i(t)$. Average power can be calculated by using $P_{av} = \frac{V_0 I_0}{2}\cos(\phi)$, where p.f. $= \cos\phi$ is the power factor for the circuit.

Quiz

1. Redo Example 10-5 with a single capacitor connected in parallel to a voltage source with $v(t) = V_0\cos(\omega t)$, assuming that the current lags the voltage by 90°. What is the average power?
2. Referring to Problem 1, what is the instantaneous power?
3. Consider the circuit shown in Fig. 10-7. Insert a corrective capacitor in *series* with the resistor and inductor. What value should C have so that the power factor is a maximum?
4. Using the values from Example 10-1, what should the capacitance be to maximize the power factor if the capacitor is connected in series?
5. Consider the circuit shown in Fig. 10-12. The effective voltage of the power source is 220 V. The two loads are rated at 200 W, 0.8 p.f. lagging and 200 W, 0.8 p.f. leading, respectively. Determine the effective currents delivered to each load.

CHAPTER 11

Transformers

A transformer is a circuit consisting of two or more inductors that are magnetically coupled. They can be used to step currents and voltages up or down. Suppose that one inductor is in parallel with a current source used to create a voltage across it by driving a current through it, as shown in Fig. 11-1.

A voltage will result in the inductor L_1 due to the current source since

$$v_1 = L_1 \frac{di_1}{dt} \tag{11.1}$$

By recalling that for two inductors in proximity the mutual inductance will cause or induce a voltage across the second inductor L_2 via

$$v_2 = L_2 \frac{di_2}{dt^2} \pm M \frac{di_1}{dt} \tag{11.2}$$

we are reminded that the first inductor can be used to induce a voltage across the second inductor.

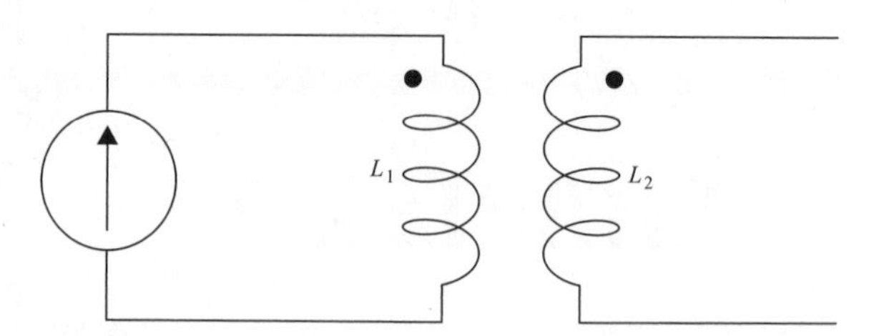

Fig. 11-1 Two inductors make up a transformer circuit.

The Dot Convention

Circuit diagrams with transformers often have dots indicated near the inductors. The dot indicates that a current flowing into the dot location will result in an added flux. We will illustrate how this is used in an example.

Consider two inductors L_1 and L_2 in a transformer with number of windings N_1 and N_2, respectively. The ratio of the winding numbers is

$$a = \frac{N_1}{N_2} \tag{11.3}$$

Then the currents flowing through each inductor will be related by

$$i_2 = \pm a i_1 \tag{11.4}$$

If i_1 flows into the dot reference of its inductor and i_2 flows out of the dot reference of its inductor, we take the + sign. On the other hand, if i_2 also flows into the dot reference of its inductor, we take the − sign.

Using phasors, the ratio of the voltage to current is called the reflective impedance and it satisfies

$$Z_r = \frac{V_1}{I_1} = \frac{aV_2}{(1/a)I_2} = a^2\frac{V_2}{I_2} = a^2 Z_2 \tag{11.5}$$

We illustrate this with an example.

EXAMPLE 11-1

Consider the circuit shown in Fig. 11-2. Find the two currents $i_1(t)$ and $i_2(t)$ if $v_s(t) = 100\cos 4t$.

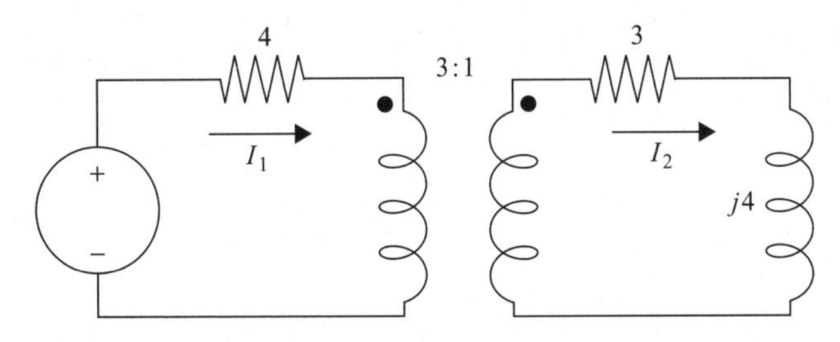

Fig. 11-2 The transformer circuit solved in Example 11-1.

SOLUTION

First, using the winding ratio given in the figure as 3:1 we have

$$3^2(3) = 27\ \Omega$$
$$3^2(j4) = j36$$

Hence we can find the current $i_1(t)$ from the circuit shown in Fig. 11-3.

Adding up the resistors in series, we have $27 + 4 = 31\ \Omega$. Therefore, applying KVL to the impedances we have

$$-V_s + 31I_1 + j36I_1 = 0$$
$$\Rightarrow$$
$$I_1 = \frac{V_s}{31 + j36} = \frac{100\angle 0^\circ}{31 + j36}$$

Notice that the current I_1 is referenced into the dotted terminal, while the current I_2 is referenced out of the dotted terminal. This means that $I_2 = +NI_1$ where N is the turn ratio. Let's compute the magnitude of the denominator in

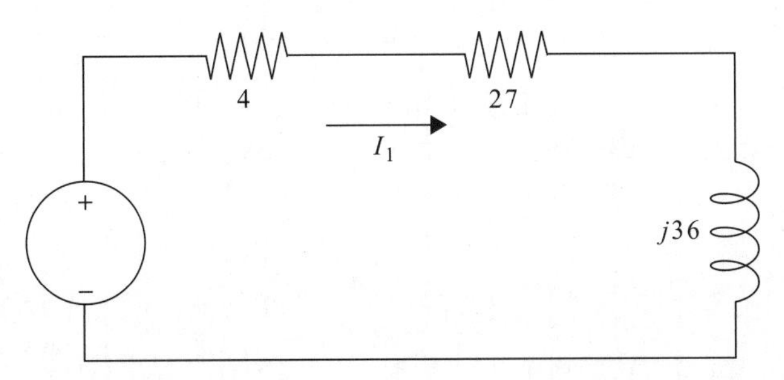

Fig. 11-3 The transformed version of the circuit shown in Fig. 11-2.

the expression for I_1. We have

$$|31 + j36| = \sqrt{31^2 + 36^2} = 47.51$$

The phase angle is

$$\phi = \tan^{-1}\left(\frac{36}{31}\right) = 49.3^\circ$$

Hence

$$I_1 = \frac{100\angle 0^\circ}{31 + j36} = \frac{100\angle 0^\circ}{47.51\angle 49.3^\circ} = 2.11\angle -49.3^\circ$$

Transforming to the time domain

$$i_1(t) = 2.11\cos(4t - 49.3^\circ)$$

The current I_2 is given by $I_2 = 3I_1 = 3(2.11)\angle -49.3^\circ = 6.33\angle -49.3^\circ$. In the time domain

$$i_2(t) = 6.33\cos(4t - 49.3^\circ)$$

Summary

When two circuits containing inductors are brought together, a current flowing in one inductor will induce a current to flow in the second inductor. We can describe this mathematically by writing

$$v_2 = L_2\frac{di_2}{dt^2} \pm M\frac{di_1}{dt}$$

Here we refer to L as the self-inductance for inductor 2, while M is the mutual inductance describing the linkage between the two circuits.

Quiz

1. Consider the circuit shown in Fig. 11-4. If $i_1(t) = 2\cos(100t)$, what is I_2?

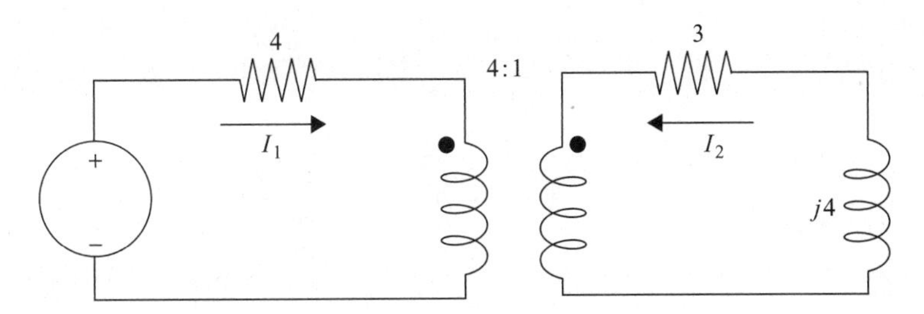

Fig. 11-4 A transformer circuit for quiz problems.

2. If the circuit shown in Fig. 11-4 is transformed similarly to the one in Example 11-1, what is the impedance of the resulting circuit?
3. If $v_s = 200 \cos 100t$, find $i_1(t)$.

CHAPTER 12

Three-Phase Circuits

A *three-phase circuit* is one that consists of three sinusoidal voltages or currents. This type of circuit is common in power generation, where an ac generator produces the three voltages. The three voltages are typically denoted by $v_a(t)$, $v_b(t)$, and $v_c(t)$. They have the same amplitude and frequency, but different phase angles

$$v_a(t) = \sqrt{2}V_{\text{eff}} \cos(\omega t + \alpha)$$
$$v_b(t) = \sqrt{2}V_{\text{eff}} \cos(\omega t + \beta)$$
$$v_c(t) = \sqrt{2}V_{\text{eff}} \cos(\omega t + \gamma)$$

where α, β, and γ are the three phase angles. So we see where the name three-phase circuit originates. The *phase sequence* is determined by finding how the voltages lead or lag each other or the order in which the voltages reach their peak values. If the order in which they reach their peak values is $v_a \rightarrow v_b \rightarrow v_c$, meaning that v_a peaks first followed by v_b etc., we say that the phase sequence is *positive.* If the three voltages peak in any other order, we call it a negative phase sequence. This idea is illustrated in Fig. 12-1.

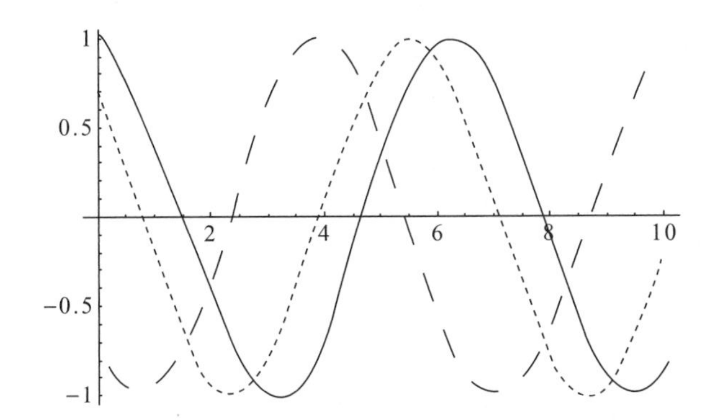

Fig. 12-1 Three voltages v_a (dashed line), v_b (dotted line), and v_c (solid line). The dashed line peaks earlier in time than the dotted line, which peaks earlier than the solid line. v_a leads v_b which leads v_c, so this is a positive sequence.

Balanced Sequences

In a three-phase circuit, the set of voltages or currents are said to be *balanced* if the difference between each phase angle is 120°.

EXAMPLE 12-1
A phase sequence is balanced. If $V_A = 170\angle 30^\circ$ and $V_C = 170\angle -90^\circ$, what is V_B?

SOLUTION
V_C lags V_A by 120°. Therefore

$$V_B = 170\angle 30^\circ + 120^\circ = 170\angle 150^\circ$$

V_B leads V_A by 120°, so the sequence is *BAC*. This is a negative sequence.

EXAMPLE 12-2
A balanced phase sequence has $V_B = 10\angle 10^\circ$ and $V_C = 10\angle 250^\circ$. What is V_A?

SOLUTION
Notice that 250° is the same as -110°, therefore V_B leads V_C by 120°. We get the phase angle for V_A from

$$\theta_A = \theta_B + 120^\circ$$

So

$$V_A = 10\angle 130^\circ$$

This is a positive sequence since V_A leads V_B.

Y Loads

If a load of impedances is connected in a Y configuration we denote the total impedance by Z_Y. Three line currents will be connecting the load to the power source. The currents are found from

$$I_A = \frac{V_A}{Z_Y}, \quad I_B = \frac{V_B}{Z_Y}, \quad I_C = \frac{V_C}{Z_Y} \tag{12.1}$$

If the phase sequence of the voltage source is ABC, then the angle of I_A will be greater than the angle of I_B by 120°, and the angle of I_B will be greater than the angle of I_c by 120°. The three line currents have the same amplitudes but different phase angles.

EXAMPLE 12-3
A balanced Y load of impedances connected to a three-phase source has $Z_Y = 100\angle 30^\circ\ \Omega$. If $V_B = 270\angle 20^\circ$, find the phasor line currents. The phase sequence is positive.

SOLUTION
First we find I_B by Ohm's law

$$I_B = \frac{V_B}{Z_Y} = \frac{270\angle 20^\circ}{100\angle 30^\circ} = 2.7\angle -10^\circ$$

With a positive phase sequence, we know that the angle of I_A is 120° larger than the angle of I_B. So

$$\theta_A = \theta_B + 120^\circ = -10^\circ + 120^\circ = 110^\circ$$

So

$$I_A = 2.7\angle 110^\circ$$

Now we subtract 120° to get the angle of I_c. That is

$$\theta_C = \theta_B - 120^\circ = -130^\circ$$

Hence

$$I_C = 2.7\angle{-130^\circ}$$

Summary

Three-phase circuits are often used in power generation.

$$v_a(t) = \sqrt{2}V_{\text{eff}} \cos(\omega t + \alpha)$$
$$v_b(t) = \sqrt{2}V_{\text{eff}} \cos(\omega t + \beta)$$
$$v_c(t) = \sqrt{2}V_{\text{eff}} \cos(\omega t + \gamma)$$

The *phase sequence* is determined by finding how the voltages lead or lag each other or the order in which the voltages reach their peak values. If the order in which they reach their peak values is $v_a \to v_b \to v_c$ meaning that v_a peaks first followed by v_b etc., we say that the phase sequence is *positive.*

Quiz

1. Consider a balanced three-phase circuit. If $V_B = 120\angle{-20^\circ}$, and $V_C = 120\angle 100^\circ$, what is V_A? Is the sequence positive or negative?
2. A Y load with impedance Z_Y is connected to a three-phase voltage source. Are the currents given by a, b, or c?

a) $I_A = \dfrac{V_A\angle\theta_A + 120^\circ}{Z_Y}$, $I_B = \dfrac{V_B}{Z_Y}$, $I_C = \dfrac{V_C\angle\theta_A - 120^\circ}{Z_Y}$

b) $I_A = \dfrac{V_A}{Z_Y}$, $I_B = \dfrac{V_B}{Z_Y}$, $I_C = \dfrac{V_C}{Z_Y}$

c) $I_A = \dfrac{V_A\angle\theta_B - 120^\circ}{Z_Y}$, $I_B = \dfrac{V_B}{Z_Y}$, $I_C = \dfrac{V_C\angle\theta_B + 120^\circ}{Z_Y}$

CHAPTER 13

Network Analysis Using Laplace Transforms

The *Laplace transform* is a mathematical tool that can be used to simplify circuit analysis. Although, at first, we seem to be adding mathematical complexity, the Laplace transform actually makes analysis easier in many cases—it transforms differential and integral equations into algebraic ones. It does this by using the fact that the exponential function is easy to differentiate and integrate. Recall that

$$\frac{d}{dt}e^{at} = ae^{at} \tag{13.1}$$

Notice that we can think of differentiation in this case as multiplication, by the constant present in the exponential. Integration is the inverse operation—in that case we divide

$$\int e^{at}\,dt = \frac{1}{a}e^{at} + C \tag{13.2}$$

where C is the constant of integration. As we will quickly see, we cam exploit these properties of the exponential function when working with Laplace transforms.

The Laplace Transform

Let $f(t)$ be some function of time. The Laplace transform of $f(t)$ is

$$\ell\{f(t)\} = F(s) = \int_0^\infty e^{-st} f(t)\, dt \tag{13.3}$$

We take s to be a complex number. Hence, we can write $s = \sigma + j\omega$, where σ is the real part of s and ω is the imaginary part of s. We refer to the function $f(t)$ as being in the *time domain* while the function $F(s)$ is in the *s domain.*

The Laplace transform is linear, that is,

$$\ell\{af(t) + bg(t)\} = a\,\ell\{f(t)\} + b\,\ell\{g(t)\} \tag{13.4}$$

where a and b are constants. The Laplace transform is quite general and can be applied to a wide variety of functions. We won't be too worried about actually calculating Laplace transforms directly. Instead we will just list some Laplace transforms of functions that are commonly encountered in circuit analysis and show how to work with them. However, let's calculate a few examples explicitly.

EXAMPLE 13-1
Find the Laplace transform of the constant function $f(t) = c$

SOLUTION
Looking at the definition (13.3) we find

$$\ell\{c\} = F(s) = \int_0^\infty ce^{-st}\, dt = -\frac{c}{s}e^{-st}\Big|_0^\infty = \frac{c}{s}$$

Hence, the Laplace transform of $f(t) = 1$ is

$$\ell\{1\} = \frac{1}{s} \tag{13.5}$$

EXAMPLE 13-2
What is the Laplace transform of $f(t) = e^{at}$?

SOLUTION

Applying (13.3) and taking $s > a$ we have

$$\ell\{e^{at}\} = F(s) = \int_0^\infty e^{at} e^{-st}\, dt = \int_0^\infty e^{(a-s)t}\, dt = \frac{1}{a-s} e^{(a-s)t} \Big|_0^\infty = \frac{1}{s-a}$$

In Table 13-1, we list the Laplace transforms of some elementary functions that are frequently seen in electrical engineering.

When a function $f(t)$ has a given Laplace transform $F(s)$, we say that we have a *Laplace transform pair* and write

$$f(t) \Leftrightarrow F(s)$$

Table 13-1 Common Laplace transforms.

$f(t)$	$F(s)$
$u(t)$	$\frac{1}{s}$
$t^n, \quad n = 1, 2, 3, \ldots$	$\frac{n!}{s^{n+1}}, s > 0$
e^{at}	$\frac{1}{s-a}, \ s > a$
$\cos \omega t$	$\frac{s}{s^2 + \omega^2}, s > 0$
$\sin \omega t$	$\frac{\omega}{s^2 + \omega^2}, \ s > 0$
$\cosh at$	$\frac{a}{s^2 - a^2}, \ s > \lvert a \rvert$
$\sinh at$	$\frac{s}{s^2 - a^2}, \ s > \lvert a \rvert$
$\delta(t)$	1
$e^{-at} u(t)$	$\frac{1}{s+a}$
$e^{-at} f(t)$	$F(s+a)$
$tf(t)$	$-\frac{d}{ds} F(s)$
$f(at)$	$\frac{1}{a} F\left(\frac{s}{a}\right)$
$\frac{df}{dt}$	$sF(s) - f(0)$
$\frac{d^2 f}{dt^2}$	$s^2 F(s) - sf(0) - f'(0)$
$\int_0^t f(\tau)\, d\tau$	$\frac{1}{s} F(s)$

In Table 13-1, notice that the Laplace transform turns differentiation in the time domain into multiplication by s in the s domain, while integration in the time domain turns into division by s in the s domain.

EXAMPLE 13-3
What is the Laplace transform of $f(t) = e^{-2t} \cos t$?

SOLUTION
We could compute the transform of this function directly by using (13.3), but that would turn into a tedious exercise of integration by parts. Instead we look at Table 13-1 and notice that

$$\ell\{\cos \omega t\} = \frac{s}{s^2 + \omega^2}$$

We also see, according to Table 13-1, that the Laplace transform of $e^{-at} f(t)$ is given by $F(s + a)$. So we set $\omega = 1$ and have

$$\ell\{\cos t\} = \frac{s}{s^2 + 1}$$

Hence

$$F(s) = \ell\{e^{-2t} \cos t\} = \frac{s - 2}{(s - 2)^2 + 1}$$

EXAMPLE 13-4
What is the Laplace transform of $f(t) = 7t^3$?

SOLUTION
First we use the linearity of the Laplace transform to write

$$F(s) = \ell\{7t^3\} = 7\ell\{t^3\}$$

Now, according to Table 13-1 the Laplace transform of t^n is

$$\frac{n!}{s^{n+1}}, s > 0$$

So we find

$$F(s) = 7\ell\{t^3\} = 7\frac{3!}{s^{3+1}} = 7\frac{6}{s^4} = \frac{42}{s^4}$$

EXAMPLE 13-5

What is the Laplace transform of $f(t) = 5\sin 2t - 2e^{-2t}\cos 4t$?

SOLUTION

We begin by using the linearity of the Laplace transform to break the calculation into two parts

$$F(s) = \ell\{f(t)\} = \ell\{5\sin 2t - 2e^{-2t}\cos 4t\} = 5\ell\{\sin 2t\} - 2\ell\{e^{-2t}\cos 4t\}$$

Looking at the first term and checking Table 13-1, we see that

$$\ell\{\sin 2t\} = \frac{2}{s^2+4}$$

For the second term, we use the fact that the Laplace transform of $\cos\omega t$ is given by $\frac{s}{s^2+\omega^2}$, $s > 0$ together with the fact that the Laplace transform of $e^{-at}f(t)$ is $F(s+a)$ to write

$$\ell\{e^{-2t}\cos 4t\} = \frac{s-2}{(s-2)^2+16}$$

Combining our results, we find

$$\begin{aligned} F(s) &= 5\ell\{\sin 2t\} - 2\ell\{e^{-2t}\cos 4t\} \\ &= 5\frac{2}{s^2+4} - 2\frac{s-2}{(s-2)^2+16} = \frac{10}{s^2+4} - \frac{2s-4}{(s-2)^2+16} \end{aligned}$$

Exponential Order

We say that a function $f(t)$ is of *exponential order* if we can find constants M and a such that

$$|f(t)| \le Me^{at} \tag{13.6}$$

for $t > T$. In electrical engineering, we are typically interested in finding out if a function is of exponential order as $t \to \infty$, and we say that the function $f(t)$ is of exponential order if $\lim\limits_{t\to\infty} |f(t)e^{-at}| = 0$.

EXAMPLE 13-6

Is the function $f(t) = \cos 2t$ of exponential order?

SOLUTION

In this case we have

$$\lim_{t\to\infty} |f(t)e^{-at}| = \lim_{t\to\infty} |\cos 2t\, e^{-at}|$$

Now $\cos 2t$ oscillates between ± 1, so all we need to do is find an a such that $\lim_{t\to\infty} |\cos 2t\, e^{-at}| = 0$. Clearly, this is true for any $a > 0$; therefore, $\cos 2t$ is of exponential order.

The Inverse Laplace Transform

When using the Laplace transform to solve equations, we complete the following steps

- Transform every term in the equation into the s domain.
- Do algebraic manipulations to find a solution in the s domain.
- Invert the solution to find a solution to the original equation in the time domain.

The last step, transforming back to the time domain, is known as the *inverse Laplace transform.* If the equation is very simple, we might get lucky and get a solution in the s domain that exactly matches the listing in Table 13-1. In that case we just read off the answer. Otherwise some algebraic manipulation will be necessary. We will denote the inverse Laplace transform by ℓ^{-1}.

EXAMPLE 13-7

Find the inverse Laplace transform of $F(s) = \dfrac{1}{s^3}$.

SOLUTION

Since we have the relation

$$\ell\{t^n\} = \frac{n!}{s^{n+1}}$$

We see that for $F(s) = \frac{1}{s^3} = \frac{1}{s^{2+1}}$, we can take $n = 2$. We are missing the $2! = 2$ term in the numerator, so the inverse Laplace transform must be

$$\ell^{-1}\left\{\frac{1}{s^3}\right\} = \ell^{-1}\left\{\left(\frac{1}{2}\right)\frac{2}{s^3}\right\} = \left(\frac{1}{2}\right)\ell^{-1}\left\{\frac{2}{s^3}\right\} = \frac{1}{2}t^2$$

In many cases, the solution to a problem will generate an expression that cannot be readily inverted. A useful tool to apply in that case is the method of *partial fractions*. This is best illustrated by example.

EXAMPLE 13-8

Find the inverse Laplace transform of

$$F(s) = \frac{3s^2 + 2}{(s+1)(s-2)(s-4)}$$

SOLUTION

The first step in the method of partial fractions is to write

$$\frac{3s^2 + 2}{(s+1)(s-2)(s-4)} = \frac{A}{s+1} + \frac{B}{s-2} + \frac{C}{s-4}$$

Hence, we need to determine the constants A, B, and C. We begin by multiplying both sides by $(s+1)(s-2)(s-4)$, giving

$$3s^2 + 2 = A(s-2)(s-4) + B(s+1)(s-4) + C(s+1)(s-2)$$

Now let's eliminate each of the variables in turn. If we let $s = -1$, we eliminate B and C and obtain

$$\begin{aligned}
&3(-1)^2 + 2 = A(-1-2)(-1-4), \Rightarrow \\
&5 = 15A, \\
&A = 1/3
\end{aligned}$$

Now eliminate A and C by letting $s = 2$. Then we find

$$\begin{aligned}
&3(2)^2 + 2 = B(2+1)(2-4), \Rightarrow \\
&14 = -6B, \\
&B = -7/3
\end{aligned}$$

Finally, we eliminate A and B by letting $s = 4$.

$$3(4)^2 + 2 = C(4 + 1)(4 - 2), \Rightarrow$$
$$50 = 10C, \Rightarrow$$
$$C = 5$$

So we have found that

$$\frac{3s^2 + 2}{(s + 1)(s - 2)(s - 4)} = \frac{1}{3}\left(\frac{1}{s + 1}\right) - \frac{7}{3}\left(\frac{1}{s - 2}\right) + \frac{5}{s - 4}$$

Using the fact that the Laplace transform of e^{at} is $1/(s - a)$, we conclude that the inverse Laplace transform of this expression is

$$\frac{1}{3}e^{-t} - \frac{7}{3}e^{2t} + 5e^{4t}$$

EXAMPLE 13-9

Find the inverse Laplace transform of

$$\frac{s - 4}{(s + 2)(s^2 + 4)}$$

SOLUTION

Since there is a quadratic in the denominator, we write our partial fraction expansion as

$$F(s) = \frac{s - 4}{(s + 2)(s^2 + 4)} = \frac{A}{s + 2} + \frac{Bs + C}{s^2 + 4} \tag{13.7}$$

Let's cross multiply by $(s + 2)(s^2 + 4)$ to give

$$s - 4 = A(s^2 + 4) + (Bs + C)(s + 2) \tag{13.8}$$

Setting $s = -2$ gives

$$-2 - 4 = A((-2)^2 + 4), \Rightarrow$$

$$-6 = 8A \text{ or } A = -3/4$$

Now we can eliminate B in (13.8) by setting $s = 0$. This gives

$$-4 = -3 + (C)(2), \Rightarrow C = -1/2$$

Finally, we can solve for B by letting s be any number. Let's pick $s = 1$. Then (13.8) becomes

$$1 - 4 = A(1^2 + 4) + (B + C)(1 + 2), \Rightarrow$$
$$\frac{-3 - 5A - 3C}{3} = B$$

Putting in $A = -3/4$ and $C = -1/2$ we find $B = 3/4$. With these values (13.7) becomes

$$F(s) = \left(-\frac{3}{4}\right)\frac{1}{s+2} + \left(\frac{3}{4}\right)\frac{s}{s^2+4} - \left(\frac{1}{2}\right)\frac{1}{s^2+4}$$
$$= \left(-\frac{3}{4}\right)\frac{1}{s+2} + \left(\frac{3}{4}\right)\frac{s}{s^2+4} - \left(\frac{1}{4}\right)\frac{2}{s^2+4}$$

We can find the inverse Laplace transform by looking at Table 13-1. Note that the following Laplace transform pairs are useful in this case

$$e^{at} \Leftrightarrow \frac{1}{s-a}, \qquad \cos\omega t \Leftrightarrow \frac{s}{s^2+\omega^2}, \qquad \sin\omega t \Leftrightarrow \frac{\omega}{s^2+\omega^2}$$

Hence, the inverse Laplace transform is

$$f(t) = -\frac{3}{4}e^{-2t} + \frac{3}{4}\cos 2t - \frac{1}{4}\sin 2t$$

Analyzing Circuits Using Laplace Transforms

Now that we have a working knowledge of how to calculate Laplace transforms and how to invert them to get back a function in the time domain, let's see how they can be applied to circuit analysis. Two important Laplace transform pairs that will help us solve RCL circuits will be

$$\frac{df}{dt} \Leftrightarrow sF(s) - f(0)$$
$$\frac{d^2f}{dt^2} \Leftrightarrow s^2F(s) - sf(0) - f'(0)$$

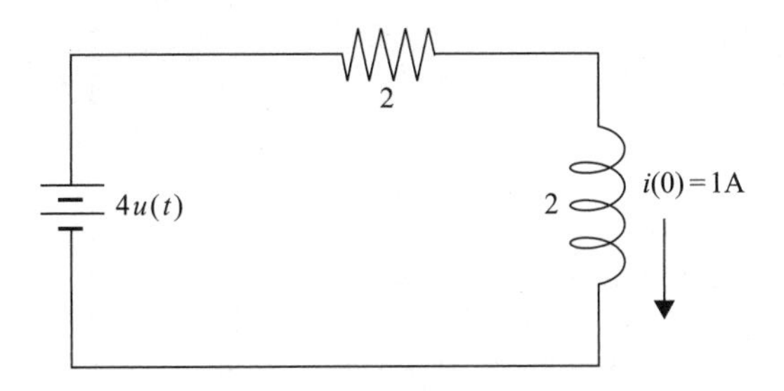

Fig. 13-1 An RL circuit.

Again, it is best to illustrate the method by example. In Example 13-10, we solve the RL circuit shown in Fig. 13-1 by using the Laplace transform.

EXAMPLE 13-10

Find the current $i(t)$ in the RL circuit shown in Fig. 13-1.

SOLUTION

Notice we have indicated that the voltage source is turned on at $t = 0$ by making it a unit step function. The equation for this circuit is

$$2\frac{di}{dt} + 2i(t) = 4u(t) \qquad (13.9)$$

Dividing by 2 to simplify we have

$$\frac{di}{dt} + i(t) = 2u(t)$$

Now we take the Laplace transform of this equation to obtain

$$I(s) + sI(s) - i(0) = \frac{2}{s} \qquad (13.10)$$

(Refer to Table 13-1 if you are not sure how we obtained each term.) Rearranging we have

$$I(s)(s+1) - 1 = \frac{2}{s}$$

Solving for $I(s)$ gives

$$I(s) = \frac{2}{s(s+1)} + \frac{1}{s+1}$$

This expression is conveniently broken up into two parts. In fact, right here we have the zero-state response and the zero-input response of the circuit. Recall that the zero-state response is due to the input source (i.e., the voltage source in this case) and not to any initial conditions (or the "initial state") of the circuit. Since the Laplace transform of the voltage source (the input) is $4/s$ and the first term on the right-hand side has an expression of this form, we recognize that the zero-state response for this circuit is given by

$$\frac{2}{s(s+1)}$$

Now recall that the zero-input response is the response of the circuit due to the initial conditions, which in this case is the initial current $i(0) = 1$ A. This is the second term

$$\frac{1}{s+1}$$

This is the response of the circuit due to the initial state without the input voltage source. Now we can compute the inverse Laplace transform to find the current as a function of time. We find that it is

$$i(t) = 2u(t) - e^{-t}u(t)$$

If you compute the Laplace transform of each term individually, you will find that the zero-state response of the circuit is

$$2(1 - e^{-t})$$

and the zero-input response is

$$e^{-t}$$

EXAMPLE 13-11

Using Laplace transform methods, find the current $i(t)$ through the inductor shown in Fig. 13-2. Then consider the natural response of the circuit and take $R = 3$, $L = 1$, and $C = 1/2$. Assume that the initial current is zero.

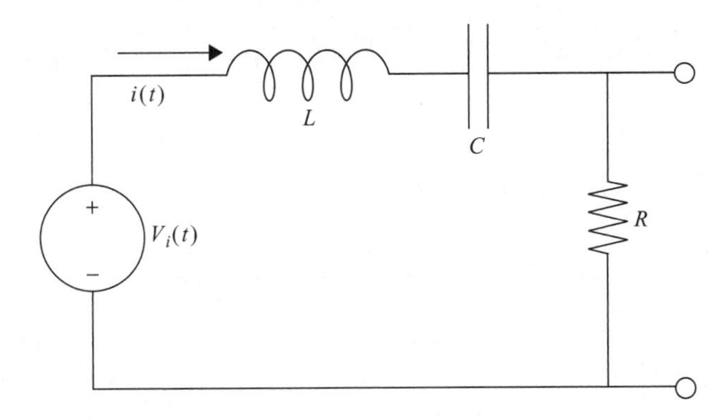

Fig. 13-2 A series RLC circuit.

SOLUTION

First, we apply KVL to the circuit. We obtain

$$v_i(t) = L\frac{di}{dt} + \frac{1}{C}\int_0^t i(\tau)\,d\tau + R\,i(t) \tag{13.11}$$

Solving this equation for the current might not be pleasant, but the Laplace transform greatly simplifies the situation. Now we take the Laplace transform of each piece. On the left-hand side, we simply have the Laplace transform of the input voltage:

$$v_i(t) \to V_i(s)$$

On the right-hand side, using Table 13-1, as a guide we find

$$L\frac{di}{dt} + \frac{1}{C}\int_0^t i(\tau)\,d\tau + Ri(t) \to LsI(s) - Li(0) + RI(s) + \frac{1}{Cs}I(s)$$

Equating this result to $V_i(s)$ and solving for $I(s)$, we find

$$I(s) = \frac{V_i(s) + L\,i(0)}{Ls + R + \frac{1}{Cs}} = \frac{s\,V_i(s) + s\,L\,i(0)}{Ls^2 + Rs + 1/C} \tag{13.12}$$

We are asked to find the *natural response* of the circuit, which means that we take $v_i(t) = \delta(t)$, the unit impulse or Dirac delta function. The Laplace transform of the unit impulse is unity so

$$V_i(s) = 1$$

Letting $R = 3, L = 1,$ and $C = 1/2$ and setting the initial current to zero gives

$$I(s) = \frac{s}{s^2 + 3s + 2} = \frac{s}{(s+1)(s+2)}$$

We will invert this result by using partial fractions. We have

$$\frac{s}{(s+1)(s+2)} = \frac{A}{s+1} + \frac{B}{s+2}$$

Multiplying both sides by $(s+1)(s+2)$ gives

$$s = A(s+2) + B(s+1)$$

If we let $s = -1$, we can eliminate B and this equation tells us that $A = -1$. On the other hand, if we let $s = -2$, then we eliminate A and find that $B = 2$. Therefore

$$I(s) = \frac{-1}{s+1} + \frac{2}{s+2}$$

Referring to Table 13-1, we see that we have the Laplace transform pair

$$e^{-at}u(t) \Leftrightarrow \frac{1}{s+a}$$

Hence, the current as a function of time is

$$i(t) = 2e^{-2t} - e^{-t}$$

This is shown in the plot in Fig. 13-3.

Convolution

The *convolution theorem for Laplace transforms* is one of the most important tools in the electrical engineer's toolbox. In the time domain, convolution can

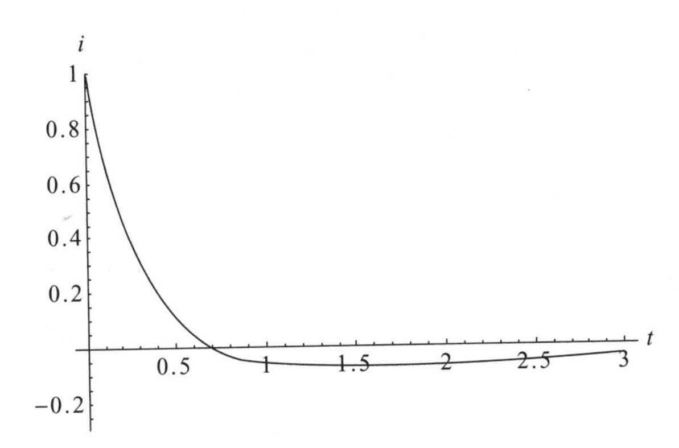

Fig. 13-3 A plot of the current $i(t) = 2e^{-2t} - e^{-t}$, which was the solution found in Example 13-11.

be a complicated integral. The definition of convolution is the following. We multiply one function by the other with the argument time shifted and then integrate

$$f * g = \int_0^t f(u)g(t-u)\,du \tag{13.13}$$

The beauty of the Laplace transform is that in the s domain, convolution is transformed from a complicated integral into a simple multiplication operation. That is, the convolution defined in (13.13) becomes

$$\ell\{f * g\} = F(s)G(s) \tag{13.14}$$

So we have a simple algorithm that we can use to determine the convolution of two functions in the time domain. This is done by applying the following steps

- Find the Laplace transforms of $f(t)$ and $g(t)$ which are $F(s)$ and $G(s)$.
- Multiply them together.
- Invert the result to get $f(t) * g(t)$.

Of course, the inversion process may not always be simple in practice, but multiplication and inversion, in general, are easier to carry out than integration.

EXAMPLE 13-12

Use the convolution theorem to solve the integral equation

$$f(t) = 2\cos t - \int_0^t \sin(u) f(t-u)\,du$$

SOLUTION

Taking the Laplace transform of both sides and using (13.14), we obtain

$$F(s) = 2\frac{s}{s^2+1} - \left(\frac{1}{s^2+1}\right) F(s)$$

Grouping together terms multiplying $F(s)$ this becomes

$$F(s)\left(1 + \frac{1}{s^2+1}\right) = 2\frac{s}{s^2+1}$$

but notice that

$$1 + \frac{1}{s^2+1} = \frac{s^2+1}{s^2+1} + \frac{1}{s^2+1} = \frac{s^2+2}{s^2+1}$$

and so we have

$$F(s)\left(\frac{s^2+2}{s^2+1}\right) = 2\frac{s}{s^2+1}$$

Dividing both sides by $\left(\frac{s^2+2}{s^2+1}\right)$ gives

$$F(s) = 2\frac{s}{s^2+1}\left(\frac{s^2+1}{s^2+2}\right) = 2\frac{s}{s^2+2}$$

Taking the inverse Laplace transform of both sides gives the solution

$$f(t) = 2\cos(\sqrt{2}\,t)u(t)$$

Zero-State Response and the Network Function

Given an electric circuit we can calculate its response to a unit impulse input $\delta(t)$. This type of excitation serves to characterize the circuit itself and can be used to determine the response of the circuit to any general type of excitation. As a result the response of the circuit to the unit impulse is sometimes called the natural response or *unit impulse response*. We denote the output by $h(t)$. This is shown schematically in Fig. 13-4.

We call the Laplace transform of the natural response the *network function* $H(s)$. Once the network function for a given circuit is known, we can calculate the response of the circuit to any input by using the convolution theorem. Let $e(t)$ represent an arbitrary excitation of the circuit. Then the response is found by using convolution

$$r(t) = \int_0^t e(u)h(t-u)du \tag{13.15}$$

The convolution theorem allows us to write this as a simple multiplication in the s domain

$$R(s) = E(s)H(s) \tag{13.16}$$

EXAMPLE 13-13

It is known that the unit impulse response for a particular circuit is $h(t) = e^{-t}u(t)$. Find the response of the circuit to the excitation $e(t) = \cos 2t$.

SOLUTION

We can find the response using convolution by calculating

$$r(t) = \int_0^t \cos(2u)e^{-(t-u)}du$$

$\delta(t)$ → circuit → $h(t)$

Fig. 13-4 The natural response of a circuit is its response when the input is a unit impulse function.

However, instead of using integration by parts, the Laplace transform makes the solution of this problem much easier. First we use (13.16), together with the fact that the Laplace transform of $h(t) = e^{-t}u(t)$ is $1/(s+1)$ and the Laplace transform of $e(t) = \cos 2t$ is $s/(s^2+4)$, to write

$$R(s) = H(s)E(s) = \frac{s}{(s^2+4)(s+1)}$$

As usual, the best route available is to find the partial fraction decomposition of this expression. We write

$$\frac{s}{(s^2+4)(s+1)} = \frac{A}{s+1} + \frac{B}{s^2+4}$$

This expression can be rewritten as

$$s = A(s^2+4) + B(s+1)$$

Now we let $s = -1$ to eliminate B

$$-1 = A(1+4) = 5A, \Rightarrow A = -\frac{1}{5}$$

Next, we let $s = 0$. This gives

$$0 = -\left(\frac{1}{5}\right)(4) + B, \Rightarrow B = \frac{4}{5}$$

So we have the following result

$$R(s) = -\left(\frac{1}{5}\right)\frac{1}{s+1} + \left(\frac{4}{5}\right)\frac{1}{s^2+4}$$
$$= -\left(\frac{1}{5}\right)\frac{1}{s+1} + \left(\frac{2}{5}\right)\frac{2}{s^2+4}$$

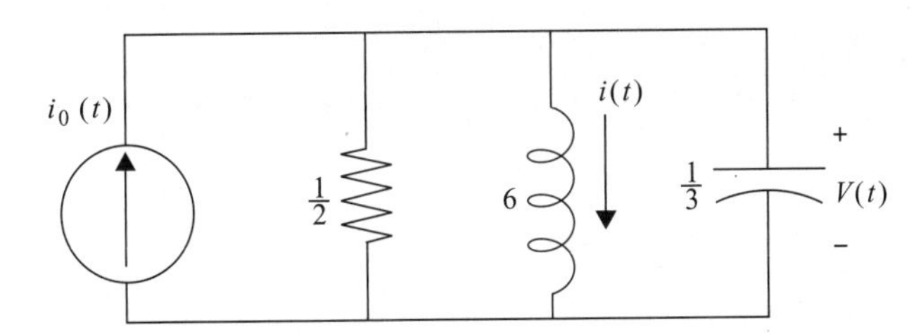

Fig. 13-5 An RLC circuit for Example 13-15.

Using Table 13-1, we invert this expression and find that the response as a function of time is

$$r(t) = -\frac{1}{5}e^{-t}u(t) + \frac{2}{5}\sin(2t)u(t)$$

We have included the unit step function to reflect the fact that the circuit is excited at time $t = 0$.

EXAMPLE 13-14

Consider the circuit shown in Fig. 13-5. If $i_0(t) = e^{-2t}u(t)$, $i(0) = 1$ A, and $v(0) = 2$ V, write a matrix equation that separates the zero-input and zero-state responses for the circuit in the s domain.

SOLUTION

Before describing the problem in terms of Laplace transforms, we use KVL and KCL to determine the equations for the unknowns $i(t)$ and $v(t)$. First, we apply KVL to the loop in the rightmost pane of the circuit. We find

$$v(t) - 6\frac{di}{dt} = 0$$

The Laplace transform of this equation is

$$\begin{aligned} &V(s) - 6(sI(s) - i(0)) = 0, \Rightarrow \\ &V(s) - 6sI(s) = -6 \end{aligned} \tag{13.17}$$

Now, we apply KCL to the top node of the inductor. We have

$$\frac{1}{3}\frac{dv}{dt} + i(t) + 2v(t) = e^{-2t}u(t)$$

Computing the Laplace transform we have

$$\frac{1}{3}(sV(s) - 2) + I(s) + 2V(s) = \frac{1}{s+2}$$

Rearranging terms gives

$$V(s)\left(\frac{1}{3}s + 2\right) + I(s) = \frac{2}{3} + \frac{1}{s+2} \tag{13.18}$$

We can combine (13.17) and (13.18) into a matrix equation

$$\begin{pmatrix} 1 & -6s \\ \frac{1}{3}s + 2 & 1 \end{pmatrix} \begin{pmatrix} V(s) \\ I(s) \end{pmatrix} = \begin{pmatrix} -6 \\ \frac{2}{3} + \frac{1}{s+2} \end{pmatrix} = \begin{pmatrix} -6 \\ \frac{2}{3} \end{pmatrix} + \begin{pmatrix} 0 \\ \frac{1}{s+2} \end{pmatrix}$$

Looking at the terms on the right-hand side, the first column vector is the zero-input response. It contains terms due only to the initial conditions

$$\begin{pmatrix} -6 \\ \frac{2}{3} \end{pmatrix}$$

The second column vector is the zero-state response. It contains terms due only to the input current source

$$\begin{pmatrix} 0 \\ \frac{1}{s+2} \end{pmatrix}$$

Poles and Zeros

We conclude the chapter with two definitions that will prove useful later when examining the stability of circuits. These are the poles and zeros of a function. Our concern in circuit analysis will be a function in the s domain. Let's say it's a rational function that can be written in the form

$$F(s) = \frac{A(s)}{B(s)}$$

The zeros of the function, as you might guess, are simply calculated by setting

$$F(s) = 0$$

Hence, we can find them by solving $A(s) = 0$. The poles of the function are the zeros of the denominator. These tell us at what values $F(s) \to \infty$ or blows up. So to find the poles we solve

$$B(s) = 0$$

EXAMPLE 13-15

In Example 13-11 we found that the current could be written in the s domain as

$$I(s) = \frac{-1}{s+1} + \frac{2}{s+2}$$

Find the poles and zeros of this function.

SOLUTION

The zeros of the function are values of s such that $I(s) = 0$. There is one zero, namely $s = 0$, in which case

$$I(0) = \frac{-1}{0+1} + \frac{2}{0+2} = -1 + 1 = 0$$

The poles are values of s for which $I(s) \to \infty$. These are the zeros in the denominator. On inspection we see that two of these are found from

$$s + 1 = 0, \Rightarrow s = -1$$
$$s + 2 = 0, \Rightarrow s = -2$$

Summary

The Laplace transform of a function of time is given by

$$\ell\{f(t)\} = F(s) = \int_0^\infty e^{-st} f(t)\, dt$$

Some frequently seen Laplace transforms include a constant

$$\ell\{1\} = \frac{1}{s}$$

an exponential

$$\ell\{e^{at}\} = \frac{1}{s-a}$$

a cosine

$$\ell\{\cos\omega t\} = \frac{s}{s^2+\omega^2}$$

and a sine function

$$\ell\{\sin\omega t\} = \frac{2}{s^2+\omega^2}$$

Laplace transforms simplify circuit analysis by letting us convert integro-differential equations into algebraic ones. Finally, we say that a function $f(t)$ is of *exponential order* if we can find constants M and a such that

$$|f(t)| \leq Me^{at}$$

Quiz

1. Find the Laplace transform of $u(t)$, the unit step function where

$$u(t) = \begin{cases} 0 & t < 0 \\ 1 & t \geq 0 \end{cases}$$

2. Find the Laplace transform of $f(t) = \cos\omega t$.
3. Compute the Laplace transform of $f(t) = e^{-t}\sin 2t$.
4. Find the Laplace transforms of $f(t) = 3t^4 + 5$ and $g(t) = 5\sin 2t$ $-3\cos 2t$.
5. Is the function $f(t) = te^{t^2}$ of exponential order?
6. Find the inverse Laplace transform of $F(s) = \dfrac{2s-3}{s^2+25}$.
7. Find the inverse Laplace transform of $F(s) = \dfrac{5s-3}{(s^2-s-2)(s+3)}$.
8. Find the inverse Laplace transform of $F(s) = \dfrac{s+1}{(s-2)(s^2+1)}$.

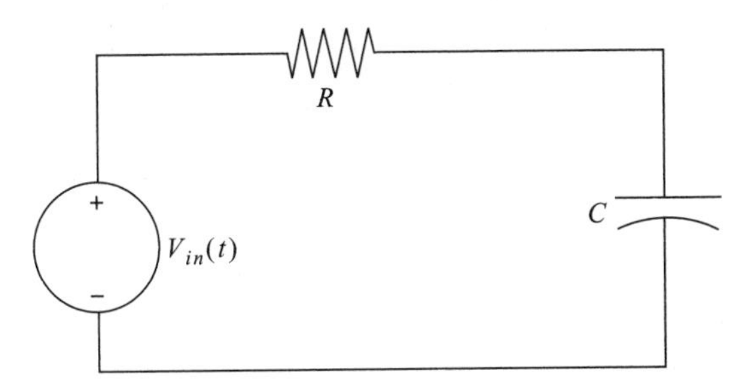

Fig. 13-6 Circuit for Problem 12.

9. Consider the circuit shown in Fig. 13-1. Suppose that instead the inductance is 4 H and the initial current is $i(0) = 3$ A. Find the zero-state and the zero-input responses as functions of time.
10. Consider the series RLC circuit shown in Fig. 13-2 and let $R = 4$, $L = 2, C = 1/2$. Suppose that the input voltage is $v_i(t) = \cos 2t$ and that the initial current is zero. Find the current $i(t)$ by using Laplace transform methods.
11. Use the convolution theorem to solve

$$f(t) = 2\cos t - \int_0^t u\, f(t-u)du$$

12. Consider the circuit shown in Fig. 13-6. Calculate the unit impulse response and network function (the voltage across the capacitor) by using the Laplace transform method.
13. Consider the circuit shown in Fig. 13-4 and suppose that $e(t) = \cos \omega t$. Find the responses $R(s)$ and $r(t)$.
14. Find the poles and zeros of $R(s) = -\left(\frac{1}{5}\right)\frac{1}{s+1} + \left(\frac{2}{5}\right)\frac{2}{s^2+4}$.

CHAPTER 14

Circuit Stability

As you might imagine, one of the most important considerations in the design of a circuit is determination of its stability. We want to find out whether a given circuit exhibits stable behavior under different conditions. By stable behavior we mean that the currents and voltages in the circuit remain bounded.

One way to determine the stability of a given circuit is to examine its transfer function. We can determine whether a circuit exhibits *impulse response stability* by looking at the behavior of $h(t)$as time increases. Formally, we want to see if

$$\lim_{t\to\infty} |h(t)| < \infty \tag{14.1}$$

Let's look at some examples.

EXAMPLE 14-1

Suppose that the transfer function for a circuit is known to be

$$H(s) = \frac{16}{s(s^2 + 8s + 16)}$$

Is the circuit impulse response stable?

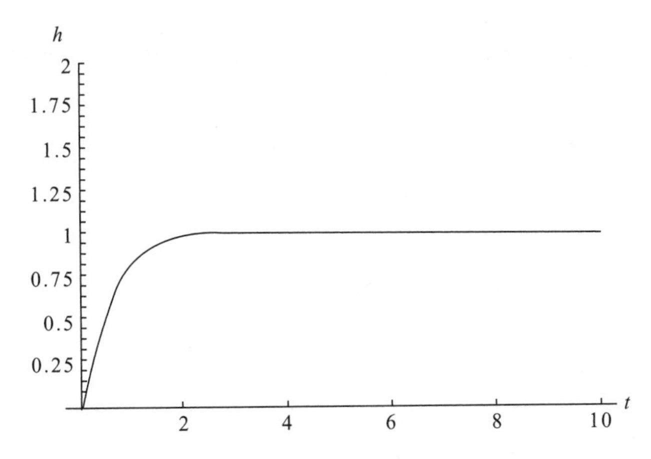

Fig. 14-1 A plot of $h(t) = 1 - e^{-4t}(4t + 1)$, which remains bounded as time increases.

SOLUTION

You can show that the inverse Laplace transform of this function, which gives us the impulse response in the time domain, is

$$h(t) = 1 - e^{-4t}(4t + 1)$$

Clearly, this circuit is stable. The limit is

$$\lim_{t\to\infty} h(t) = \lim_{t\to\infty} 1 - e^{-4t}(4t + 1) = 1$$

A plot of the function in Fig. 14-1 shows that it rapidly rises to unity, where it remains fixed for all time.

EXAMPLE 14-2

Examine the impulse response stability for a circuit with

$$H(s) = \frac{16}{s(s^2 + 2s + 4)}$$

SOLUTION

Calculating the inverse Laplace transform we find

$$h(t) = 16\left(\frac{1}{4} - \frac{1}{12}e^{-t}(\cos\sqrt{3}t + \sqrt{3}\sin\sqrt{3}t)\right)$$

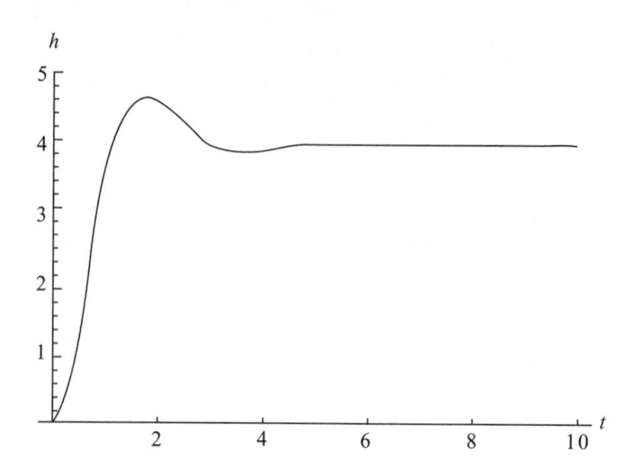

Fig. 14-2 The oscillating transfer function of Example 14-2 which has a constant steady-state value.

We find that the limit is

$$\lim_{t \to \infty} h(t) = 4$$

A plot of the function is shown in Fig. 14-2. The function steadily rises to a maximum, then decays with a tiny bit of oscillatory behavior and reaches the steady-state value.

EXAMPLE 14-3

Examine the impulse response stability for a circuit with

$$H(s) = \frac{s + 2}{(s + 2)^2 + 400}$$

SOLUTION

Inverting, we find

$$h(t) = e^{-2t} \cos 20t$$

Clearly

$$\lim_{t \to \infty} h(t) = 0$$

As Fig. 14-3 shows, this function starts off with rapid oscillations but quickly dies off to zero.

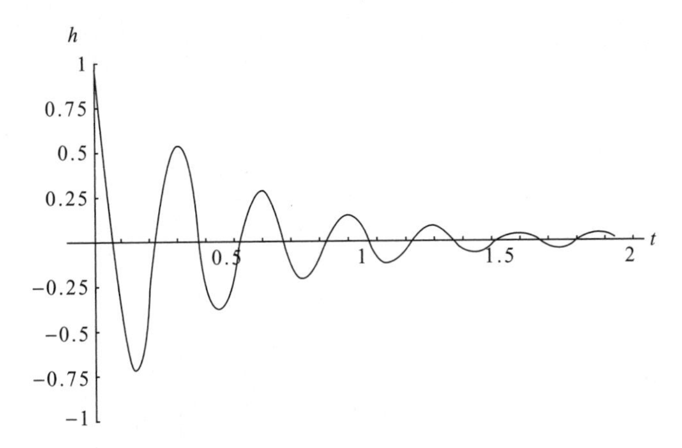

Fig. 14-3 A plot of $h(t) = e^{-2t} \cos 20t$.

Sometimes the requirements for unit impulse stability are stronger. We may require that

$$\lim_{t \to \infty} |h(t)| = 0 \tag{14.2}$$

By this criterion, only Example 14-3 is stable.

Poles and Stability

We can also get an idea about whether a circuit is impulse response stable by looking at the poles of the transfer function $H(s)$. Let's begin with a simple case. We are familiar with the Laplace transform pair

$$e^{-at} \leftrightarrow \frac{1}{s+a} \tag{14.3}$$

Provided that $a > 0$, we know that as $t \to \infty$, $h(t) = e^{-at} \to 0$. Therefore this represents an impulse response stable circuit. So if

$$H(s) = \frac{1}{s+a}$$

which has one simple pole given by

$$s + a = 0, \Rightarrow s = -a$$

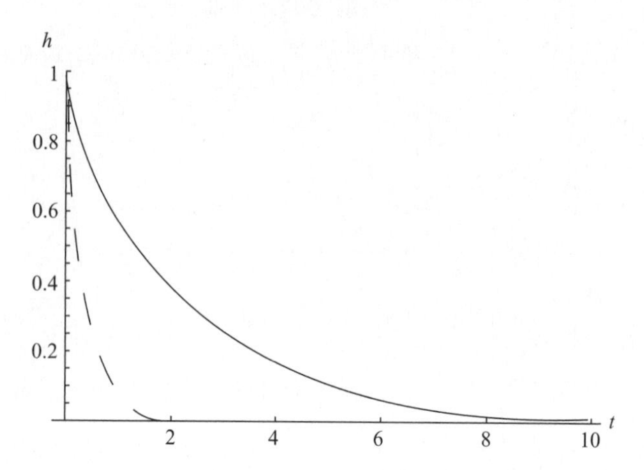

Fig. 14-4 A comparison of of $e^{-0.5t}$ (solid line) and e^{-3t}(dashed line).

If the pole is real and $a > 0$, the circuit is impulse response stable. If the pole is small, then the function will decay slowly with time. On the other hand, if the pole is large, $h(t)$ decays rapidly. This is illustrated in Fig. 14-4, where we show a plot of $e^{-0.5t}$ (solid line) and e^{-3t} (dashed line). The poles of these two functions are $s = -1/2$ and $s = -3$, respectively. In the latter case, the pole has a larger magnitude and so the transfer function decays faster.

Now let's begin to consider more complicated cases. The next complication we might imagine with poles are multiple poles of the type $s = -a$, that is

$$H(s) = \frac{1}{(s+a)^n} \tag{14.4}$$

The inverse Laplace transform of this expression is given by

$$h(t) = t^{n-1}e^{-at} \tag{14.5}$$

Since we have t raised to an integral, power multiplies by an exponential, which rises or decays very rapidly; the behavior of the exponential will dictate the behavior of the transfer function. In particular, if $a > 0$ then the transfer function will decay to zero as $t \to \infty$, and the circuit will be impulse response stable.

EXAMPLE 14-4

Consider

$$H(s) = \frac{3}{(s+2)^2}$$

and discuss its stability.

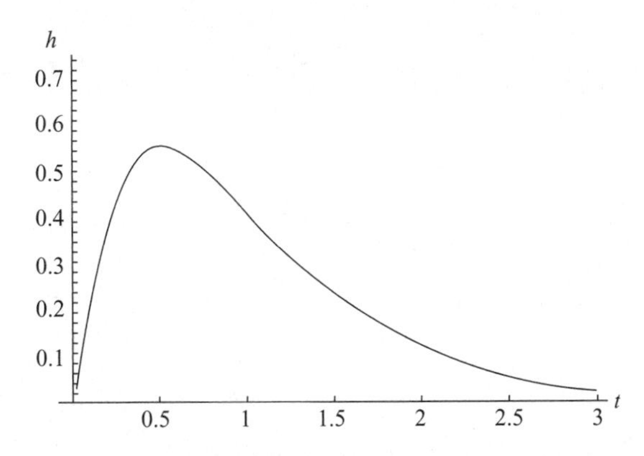

Fig. 14-5 A plot of the stable transfer function $h(t) = 3te^{-2t}$.

SOLUTION

The function has a multiple pole at

$$s = -2$$

Hence, we can see that it will be stable. The inverse Laplace transform of the function is given by

$$h(t) = 3te^{-2t}$$

A plot of the function is shown in Fig. 14-5. We see that the function steeply rises to a maximum (behavior due to the monomial term) and then it quickly decays when the exponential becomes dominant. It's easy to see that the maximum occurs at $t = 0.5$ since

$$\frac{dh}{dt} = 3e^{-2t} - 6te^{-2t} = 0$$

$$\Rightarrow t = 0.5$$

Next we consider sinusoidal functions. Clearly sinusoidal waveforms are bounded and so (14.1) is satisfied. The poles in the case of a sinusoidal transfer function will be *complex*. Let's consider the following function.

$$H(s) = \frac{1}{(s-2)(s+2)} \tag{14.6}$$

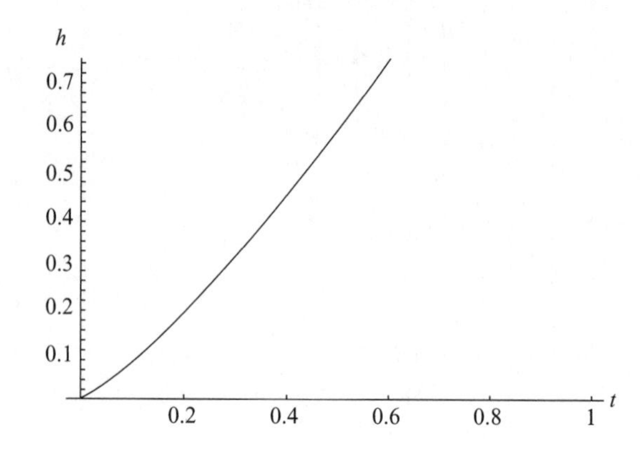

Fig. 14-6 The transfer function $h(t) = \frac{1}{4}e^{-2t}(e^{4t} - 1)$ blows up, so it's unstable.

The poles of this function are located at $s = \pm 2$. The inverse Laplace transform of this expression is

$$h(t) = \frac{1}{4}e^{-2t}(e^{4t} - 1)$$

Clearly this function blows up, as we show in Fig. 14-6.

If the roots are complex, then we get a sinusoidal function. Let's suppose that instead the poles are given by $s = \pm j2$. This corresponds to

$$H(s) = \frac{1}{(s - j2)(s + j2)} = \frac{1}{s^2 + 4} \tag{14.7}$$

As a function of time this is

$$h(t) = \frac{1}{2}\sin 2t$$

A plot of this function is shown in Fig. 14-7.

Notice that, in contrast to (14.7), the unstable transfer function in (14.6) differs by the sign, that is

$$H(s) = \frac{1}{(s - 2)(s + 2)} = \frac{1}{s^2 - 4}$$

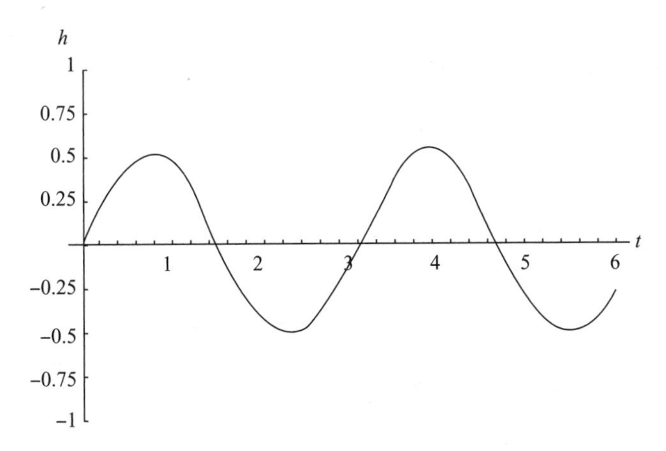

Fig. 14-7 The friendly sine function is stable since it remains bounded.

So if you see a transfer function of the form $\frac{1}{s^2-4}$ then you know it's an unstable circuit.

Next let's revisit the case of a decaying exponential multiplied by a sinusoidal. Suppose that our transfer function assumes the form

$$H(s) = \frac{s+a}{(s+a)^2+\omega^2} \tag{14.8}$$

To find the poles, we set the denominator to zero

$$(s+a)^2+\omega^2 = 0$$

So we find that the poles are

$$s = -a \pm j\omega$$

The key to stability in this case comes down to the constant a. The condition for stability is

$$a > 0, \Rightarrow \text{ impulse response stable} \tag{14.9}$$

The inverse Laplace transform of (14.8) is

$$h(t) = e^{-at}\cos\omega t \tag{14.10}$$

Looking at $h(t)$ it immediately becomes clear that if $a > 0$ the function is stable. We saw an example of this type of function in the beginning of the chapter where we met $h(t) = e^{-2t} \cos 20t$ in Fig. 14-3.

Zero-Input Response Stability

Now we consider the stability of a circuit in the zero-input response case, that is, for a circuit with initial values for currents and voltages but no sources. We can determine zero-input response stability by again looking at the poles of the transfer function $H(s)$. The first case we consider is when the poles of $H(s)$are real and negative. In that case the circuit is stable, which indicates that the currents and voltages in the circuit will decay to zero as $t \to \infty$.

EXAMPLE 14-5

Consider a series RLC circuit with $R = 2$, $C = 1/2$, and $L = 4$. If the initial voltage across the capacitor is $v_c(0) = 2$ and the initial current flowing through the capacitor is $i_c(0) = 0$, determine whether the circuit is zero input stable.

SOLUTION

As usual, KVL around the loop of a series RLC circuit gives

$$LC\frac{d^2 v_c}{dt^2} + RC\frac{dv_c}{dt} + v_c = 0$$

Taking the Laplace transform with the initial conditions specified gives the s-domain equation

$$LC(s^2 V_c(s) - sv_c(0)) + RC(sV_c(s) - v_c(0)) + V_c(s) = 0$$

Solving we have

$$V_c(s) = \frac{RCv_c(0)}{LCs^2 + (RC - LCv_c(0))s + 1}$$

Putting in the values given in the problem statement this is

$$V_c(s) = \frac{2}{s^2 - 3s + 1}$$

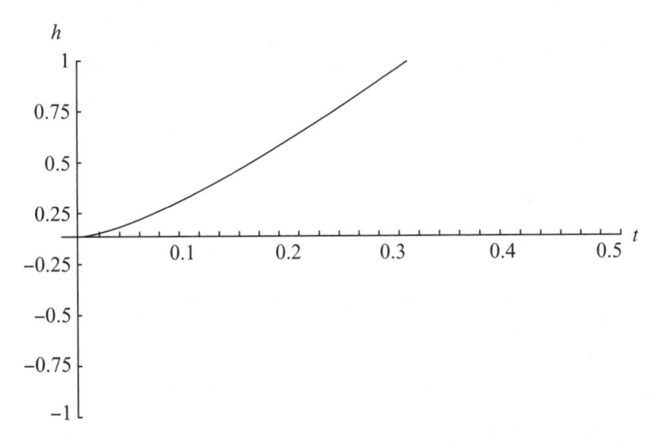

Fig. 14-8 A plot of $v_c(t)$, the voltage across the capacitor in Example 14-5. The voltage blows up so the circuit is zero input unstable.

The poles of this function are given by

$$s = \frac{3 \pm \sqrt{5}}{2}$$

These are both real and *positive,* indicating that this circuit is unstable. In fact, computing the inverse Laplace transform shows that this is a rapidly increasing function with time, as shown in Fig. 14-8.

Bounded Input-Bounded Output Stability

The final type of stability we consider is *bounded input-bounded output* or *BIBO* stability. A system is BIBO stable if a bounded input results in a bounded output. Again, we can determine stability by looking at the poles of the transfer function $H(s)$. If the poles lie in the left-hand side of the s plane, that is, they are real and negative, then the system is BIBO stable. For BIBO stability, the poles cannot lie on the imaginary ω axis because in that case we can excite the circuit with a bounded sinusoidal input

$$e(t) = A \cos \omega t$$

and if ω happens to match the natural frequency of the circuit then, even though the excitation is bounded, there will be a resonance and the response of the circuit will blow up.

EXAMPLE 14-6

Consider a series LC circuit and describe when the circuit is BIBO stable. Assume the circuit is excited with a voltage source $v(t) = A\cos\omega t$.

SOLUTION

The differential equation describing this circuit is

$$LC\frac{d^2v_c}{dt^2} + v_c = A\cos\omega t$$

Taking the Laplace transform of both sides we obtain

$$(LCs^2 + 1)V_c(s) = \frac{As}{s^2 + \omega^2}$$

Now recalling that the natural frequency is defined via

$$\omega_0^2 = \frac{1}{LC}$$

We can write the solution as

$$V_c(s) = \left(\frac{\omega_0}{s^2 + \omega_0^2}\right)\left(\frac{As}{s^2 + \omega^2}\right)$$

The transfer function is

$$H(s) = \frac{\omega_0}{s^2 + \omega_0^2}$$

with poles at

$$s = \pm j\omega_0$$

If the frequencies don't match, then the system is stable. For a trivial example suppose

$$V_c(s) = \left(\frac{4}{s^2 + 16}\right)\left(\frac{s}{s^2 + 8}\right)$$

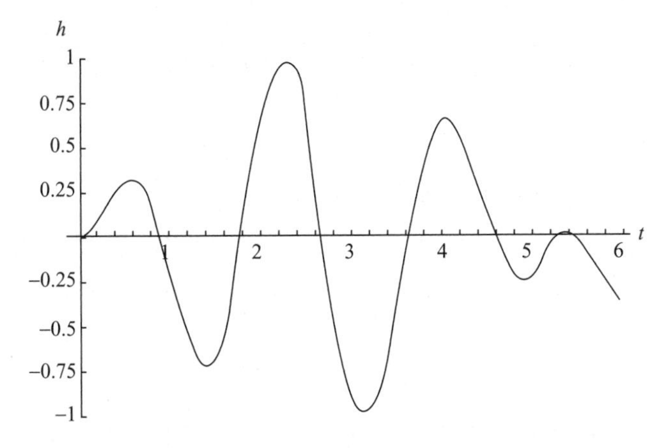

Fig. 14-9 A plot of $v_c(t) = \frac{1}{2}(\cos 2\sqrt{2}t - \cos 4t)$.

Then the voltage across the capacitor is

$$v_c(t) = \frac{1}{2}(\cos 2\sqrt{2}t - \cos 4t)$$

This is a stable voltage. A plot is shown in Fig. 14-9.

On the other hand, suppose the frequencies match (there is a resonance). Continuing the example suppose instead that

$$V_c(s) = \left(\frac{4}{s^2 + 16}\right)\left(\frac{s}{s^2 + 16}\right)$$

then

$$v_c(t) = \frac{1}{2}t \sin 4t$$

This function oscillates but grows linearly—it grows without bound. Hence, at resonance the circuit is unstable. This is shown in Fig. 14-10 where you can see the voltage across the capacitor start to grow.

Summary

Often we need to consider the stability of a circuit under different conditions. For example, do the voltages and currents remain finite as time progresses? We can characterize stability behavior by looking at the impulse response or

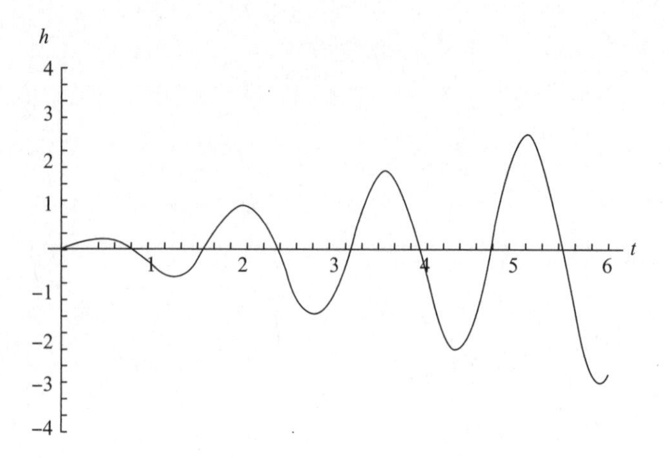

Fig. 14-10 The unstable case of resonance.

transfer function. A circuit is impulse response stable if the transfer function remains finite, that is, if $\lim_{t \to \infty} |h(t)| < \infty$. We often look at zero-input response stability (that is, no sources). This is done by looking at the poles of the transfer function $H(s)$. When the poles of $H(s)$ are real and negative, the circuit is stable; this indicates that the currents and voltages in the circuit will decay to zero as $t \to \infty$. The final type of stability we considered was *bounded input-bounded output* or *BIBO* stability. A system is BIBO stable if a bounded input results in a bounded output.

Quiz

1. Is the function $h(t) = t \sin 2t$ stable?
   In the following questions, determine the stability of the following transfer functions and find their time domain representation.
2. $H(s) = \frac{s}{s^2+16}$.
3. $H(s) = \frac{1}{s-6}$
4. $H(s) = \frac{1}{s^2}$
5. $H(s) = \frac{6}{(s+2)^2+36}$
6. Consider an LC circuit with the capacitor and inductor in series. If $C = 1/12$, $L = 4$, and the initial voltage across the capacitor is 1 V and all other initial voltages and currents are zero, determine whether the circuit is zero input stable.

CHAPTER 15

Bode Plots and Butterworth Filters

The frequency response of a system can be plotted using a logarithmic scale in the following manner. Given the frequency response $H(\omega)$, we calculate

$$|H(\omega)|_{\mathrm{dB}} = 20\ \log_{10} |H(\omega)| \tag{15.1}$$

We call this quantity the magnitude of the frequency response in decibels (dB), or sometimes we denote (15.1) by $\alpha(j\omega)$ and call it the *gain function*. A decibel is a dimensionless unit based on the ratio of two quantities. The reader is probably familiar with the use of decibels in the study of sound. In that case, we can characterize how loud a sound is by comparing the intensity I of a given sound wave to the threshold for human hearing, which we denote as I_o. We then compute the intensity of the sound in decibels as

$$I_{\mathrm{dB}} = 10 \log_{10} \left| \frac{I}{I_o} \right| \tag{15.2}$$

The multiplicative constant, 10 in this example, gives us a way to compare the relative strength differences between two quantities. That is, if the difference between I_{dB1} and I_{dB2} is 10 dB, then the intensity of I_1 is ten times the strength of I_2. Note that since the multiplicative factor in (15.1) is 20, a difference of 20 dB indicates that one signal has a magnitude 20 times as large as the other.

The log function is a useful measure of signal strength for two good reasons. First, quantities can often vary quite a bit in strength—sometimes over many orders of magnitude. By using the logarithm we can rescale that variation down to a more manageable number. One famous example where this behavior is apparent is the Richter scale used to characterize the strength of earthquakes. The details of the Richter scale don't concern us; all that is important for our purposes is that this is a logarithmic quantity. This means that each increment on the Richter scale describes an order-of-magnitude increase in strength. An earthquake that is a 7 on the Richter scale is 10 times as strong as an earthquake that is a 6. In our case, using logarithms allows us to scale down a wide range of frequencies into a small scale that can be visualized and plotted more easily. This works in a way similar to the Richter scale. In our case, when the magnitude of the frequency ω increases by a factor of 10, then $\log \omega$ increases by 1.

The second reason that using logarithms is useful is that

$$\log(AB) = \log A + \log B \tag{15.3}$$

By turning multiplication into addition, the mathematics of a problem is simplified. In engineering this can be useful when calculating the overall gain of a composite system, which could be an amplifier or filter. By using logarithms, we can simply add together the gain at each stage (in dB) to arrive at the overall gain of the system. We call each 10-to-1 change in frequency a *decade*. That is, two frequencies ω_A and ω_B are separated by one decade if

$$\omega_A = 10\omega_B \tag{15.4}$$

If one frequency is twice the other, we say they are an *octave* apart

$$\omega_A = 2\omega_B \tag{15.5}$$

Asymptotic Behavior of Functions

When analyzing the transfer function for a given system, it is important to characterize its low- and high-frequency behavior. Given a transfer function

$H(s)$, we characterize the low-frequency behavior by considering the limit

$$\lim_{s \to 0} H(s) \tag{15.6}$$

When doing the analysis in terms of frequency, we let $s \to j\omega$ and then examine the limit

$$\lim_{\omega \to 0} |H(j\omega)| \tag{15.7}$$

In the high-frequency case, in the s domain we consider the limit

$$\lim_{s \to \infty} H(s) \tag{15.8}$$

Or we let $s \to j\omega$ and examine

$$\lim_{\omega \to \infty} |H(j\omega)| \tag{15.9}$$

EXAMPLE 15-1

Determine the asymptotic behavior of the transfer function

$$H(s) = \frac{2s + 6}{s^2 - s - 12}$$

SOLUTION

First we do a bit of algebraic manipulation

$$H(s) = \frac{2s + 6}{s^2 - s - 12} = H(s) = 2\frac{s + 3}{s^2 - s - 12} = 2\frac{s + 3}{(s + 3)(s - 4)} = \frac{2}{s - 4}$$

At low frequencies

$$\lim_{s \to 0} \frac{2}{s - 4} = -\frac{1}{2}$$

To examine the function directly in frequency, we set $s \to j\omega$ and multiply top and bottom of $H(j\omega)$ by the complex conjugate

$$H(j\omega) = \frac{2}{-4 + j\omega} = \frac{2}{-4 + j\omega}\left(\frac{-4 - j\omega}{-4 - j\omega}\right) = \frac{-8 - j2\omega}{\omega^2 + 16}$$

The low-frequency behavior is

$$\lim_{\omega\to 0} H(j\omega) = \lim_{\omega\to 0} \frac{-8 - j2\omega}{\omega^2 + 16} = -\frac{8}{16} = -\frac{1}{2}$$

Now let's examine high-frequency behavior. We have

$$\lim_{s\to\infty} H(s) = \lim_{s\to\infty} \frac{2}{s-4} = 0$$

And similarly for ω. What about the behavior of the phase angle? First let's calculate it by using

$$H(j\omega) = \frac{-8 - j2\omega}{\omega^2 + 16}$$

We find

$$\theta = \tan^{-1}\left(\frac{-2\omega}{-8}\right) = \tan^{-1}\frac{\omega}{4}$$

At high frequencies

$$\theta = \lim_{\omega\to\infty} \tan^{-1}\frac{\omega}{4} = 90^\circ$$

On the other hand, at low frequencies we have

$$\theta = \lim_{\omega\to 0} \tan^{-1}\frac{\omega}{4} = \tan^{-1} 0 = 0^\circ$$

Once we understand how to characterize the low- and high-frequency behavior of the transfer function and its phase angle, we are ready to create Bode plots.

Creating Bode Plots

A Bode plot is a log–linear plot. The axes are defined in the following way:

- The horizontal axis is the logarithm of frequency ($\log_{10}\omega$)
- The vertical axis is the frequency response in decibels

In signal analysis we will plot two quantities

- The magnitude of the frequency response in decibels (15.1)
- $\theta_H(\omega)$

In our examples, we will focus on plotting the straight-line approximations to $|H(\omega)|_{dB}$ I as compared with the Bode plots of the actual function, which you can easily plot using a computational math package. Our goal here is to gain a qualitative understanding of how to generate Bode plots and what they mean.

Bode Plot Examples

The key to sketching a Bode plot is to follow these steps

- Look at very-low-frequency behavior (consider $\omega \to 0$)
- Look at very-high-frequency behavior (consider $\omega \to \infty$)
- Find the intersection with the 0 dB axis, known as the *corner frequency*

We begin with the simplest case, generating Bode plots for first-order systems.

EXAMPLE 15-2
Given that the transfer function for a given circuit is $H(s) = 1 + s$, sketch the Bode plot.

SOLUTION
We set $s = j\omega$ and obtain

$$H(\omega) = 1 + j\frac{\omega}{20}$$

We need to determine the low- and high-frequency behavior of the system. First we consider the low-frequency behavior. That is, we consider the magnitude of the frequency response $H(\omega)$ when $\omega \ll 20$. We have

$$|H(\omega)|_{\text{dB}} = 20\log_{10}\left|1 + j\frac{\omega}{20}\right| \to 20\log_{10}|1| \to 0 \text{ as } \omega \to 0$$

Next we consider the high-frequency behavior. To do this, recall that for a complex number z, the modulus is

$$|z|^2 = z\bar{z}$$

In our case, we have

$$z = 1 + j\frac{\omega}{20}, \Rightarrow \bar{z} = 1 - j\frac{\omega}{20}$$

And so we get

$$\begin{aligned} |z|^2 = \left(1 + j\frac{\omega}{20}\right)\left(1 - j\frac{\omega}{20}\right) &= 1 + j\frac{\omega}{20} - j\frac{\omega}{20} + \left(j\frac{\omega}{20}\right)\left(-j\frac{\omega}{20}\right) \\ &= 1 + \frac{\omega^2}{400} \end{aligned}$$

Now, if we consider $\omega \gg 20$, then

$$\frac{\omega^2}{400} \gg 1$$

and so we can approximate the magnitude by

$$|z|^2 = 1 + \frac{\omega^2}{400} \approx \frac{\omega^2}{400}$$

Taking the square root, $|H(\omega)| \approx \frac{\omega}{20}$. Therefore the large frequency expression for the magnitude expressed in decibels is

$$|H(\omega)|_{\text{dB}} \to 20\log_{10}\left(\frac{\omega}{20}\right) \quad \text{as } \omega \to \infty$$

This is a straight line. If we set $\omega = 20$, then we have $\log_{10}(1) = 0$, so this tells us that this line intersects the 0 dB axis at $\omega = 20$, which is the corner frequency and which we'll denote as ω_c. Combining the low- and high-frequency behavior that we have found, we have

$$|H(\omega)|_{\text{dB}} = \begin{cases} 0 & \text{for } 0 < \omega < 20 \\ 20\log_{10}\left(\frac{\omega}{20}\right) & \text{for } \omega \geq 20 \end{cases}$$

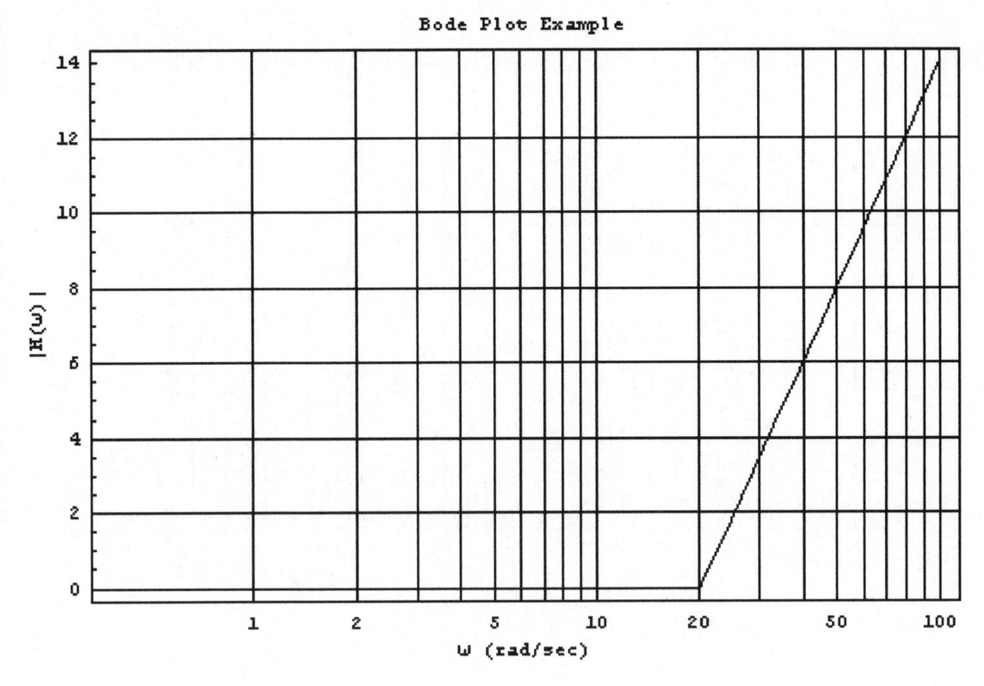

Fig. 15-1 Bode plot for Example 15-1.

The Bode plot is just a plot of this piecewise function, shown in Fig. 15-1.

Next we plot $\theta_H(\omega)$. We have

$$\theta_H(\omega) = \tan^{-1}\frac{\omega}{20}$$

The asymptotic behavior is given by

$$\theta_H(\omega) = \tan^{-1}\frac{\omega}{20} \to 0 \text{ as } \omega \to 0$$
$$\theta_H(\omega) = \tan^{-1}\frac{\omega}{20} \to \frac{\pi}{2} \text{ as } \omega \to \infty$$

A plot of $\theta_H(\omega) = \tan^{-1}\frac{\omega}{20}$ is shown in Fig. 15-2. Notice that at large frequency, $\theta_H(\omega) = \tan^{-1}\frac{\omega}{20}$ does level off at $\pi/2$.

EXAMPLE 15-3

Sketch the Bode plot for $H(\omega) = \frac{1}{1+j\frac{\omega}{10}}$.

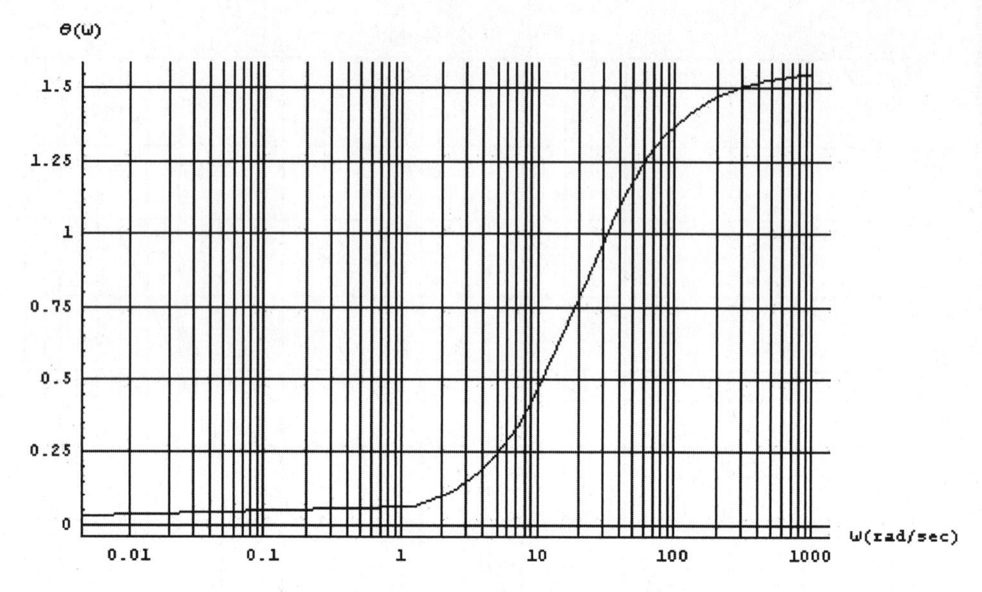

Fig. 15-2 A plot of $\theta_H(\omega) = \tan^{-1}\frac{\omega}{20}$.

SOLUTION

We follow the same procedure as before. We don't have to worry about the function being in the denominator, because when we take the log we can apply

$$\log\frac{1}{A} = -\log A$$

And so for $|H(\omega)|_{\text{dB}} = 20\log_{10}|H(\omega)|$ we have

$$|H(\omega)|_{\text{dB}} = 20\log_{10}|H(\omega)| = 20\log_{10}\left|\frac{1}{1+j\dfrac{\omega}{10}}\right| = -20\log_{10}\left|1+j\frac{\omega}{10}\right|$$

Now we can proceed using the same method we applied in the last example. We can see that as $\omega \to 0$, $|H(\omega)|_{\text{dB}} \to 0$. Therefore, the low-frequency behavior of this system, as defined for frequencies below the cutoff frequency, will be that the system remains at a constant 0 dB.

For high frequencies, we find that for $\omega \gg 10$,

$$|H(\omega)|_{\text{dB}} = 20\log_{10}|H(\omega)| = 20\log_{10}\left|\frac{1}{1+j\dfrac{\omega}{10}}\right|$$

$$\to -20\log_{10}\left|\frac{\omega}{10}\right| \text{ as } \omega \to \infty$$

Once again this is a straight line, but in this case we have a minus sign out front giving a negative slope. The corner frequency is given by $\omega = 10$, and so we have

$$|H(\omega_c)|_{\text{dB}} = 20\log_{10}|H(\omega_c)| = 20\log_{10}\left|\frac{1}{1+j\dfrac{10}{10}}\right| = 20\log_{10}\left|\frac{1}{1+j1}\right|$$

$$= -20\log_{10}|1+j| = -20\log_{10}\sqrt{2} = -3 \text{ dB}$$

Putting these results together, we see that the system response decreases with increasing frequency. This is shown in Fig. 15-3.

To characterize $\theta_H(\omega)$, we again pick up a minus sign since the function of frequency in this case is in the denominator. So we have

$$\theta_H(\omega) = -\tan^{-1}\frac{\omega}{10}$$

with asymptotic behavior given by

$$\theta_H(\omega) = -\tan^{-1}\frac{\omega}{10} \to 0 \text{ as } \omega \to 0$$

$$\theta_H(\omega) = -\tan^{-1}\frac{\omega}{10} \to -\frac{\pi}{2} \text{ as } \omega \to \infty$$

A plot of this is shown in Fig. 15-4.

EXAMPLE 15-4

For our final example, sketch the Bode plot for

$$H(\omega) = 300\frac{5+j\omega}{-\omega^2 + j11\omega_{-} + 10}$$

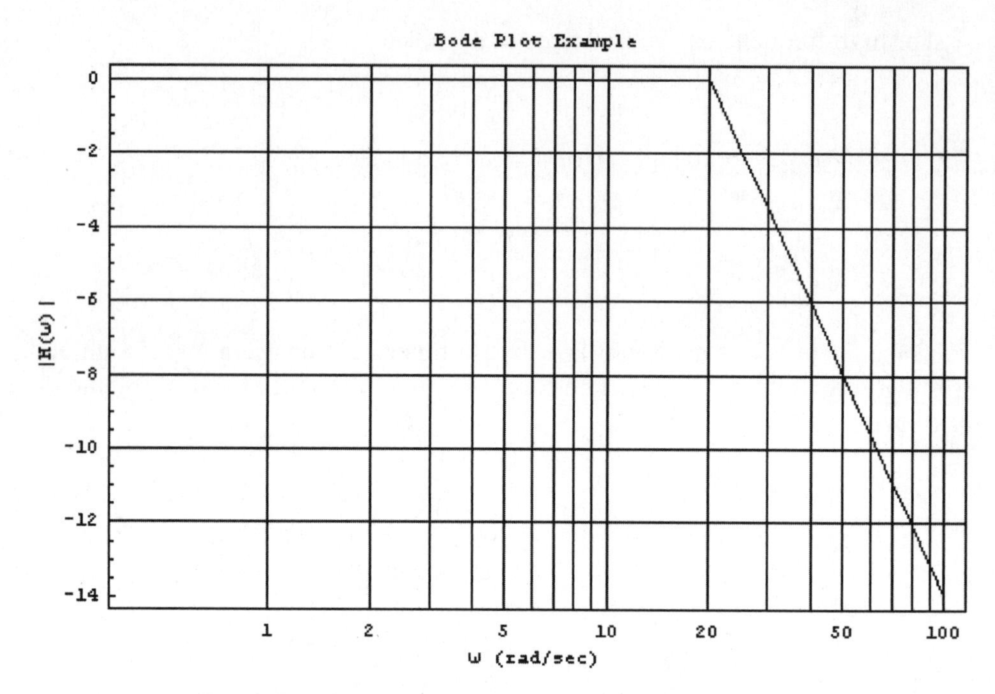

Fig. 15-3 The linearly decreasing case in Example 15-2.

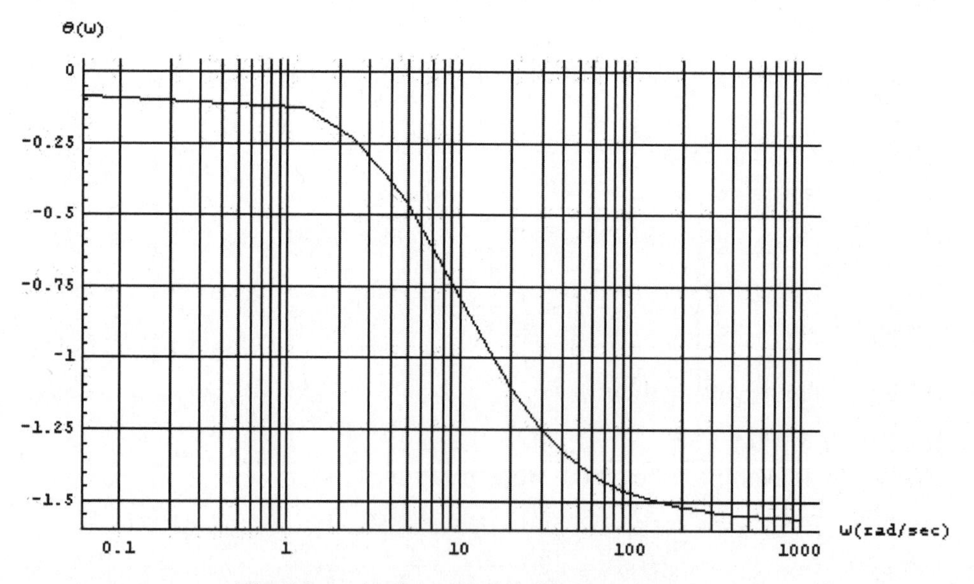

Fig. 15-4 A plot of $\theta_H(\omega)$ for Example 15-2.

SOLUTION

We begin by rewriting the transfer function in a more convenient form

$$H(\omega) = 300\frac{5 + j\omega}{-\omega^2 + j11\omega_- + 10} = 300\frac{5 + j\omega}{(1 + j\omega)(10 + j\omega)}$$

Let's factor out the 10 in the denominator and the 5 in the numerator

$$H(\omega) = 300\frac{5 + j\omega}{(1 + j\omega)(10 + j\omega)} = 300\frac{5 + j\omega}{(1 + j\omega)(10)(1 + j\omega/10)}$$
$$= 150\frac{1 + j\omega/5}{(1 + j\omega)(1 + j\omega/10)}$$

When we compute the logarithm, we can use $\log \frac{A}{BC} = \log A - \log B - \log C$. And so we have

$$|H(\omega)|_{\text{dB}} = 20\log_{10}\left|150\frac{1 + j\omega/5}{(1 + j\omega)(1 + j\omega/10)}\right| = 20\log_{10}|150|$$
$$+20\log_{10}|1+j\omega/5| - 20\log_{10}|1 + j\omega| - 20\log_{10}|1 + j\omega/10|$$

The first term is just a constant. For the other three terms, notice that there are three corner frequencies. We consider each in turn. The corner frequency for $20\log_{10}|1 + j\omega/5|$ is given by $\omega_{c1} = 1$ and we have

$$20\log_{10}|150| + 20\log_{10}|1 + j/5| - 20\log_{10}|1 + j| - 20\log_{10}|1 + j\omega/10|$$
$$= 20\log_{10}|150| + 20\log_{10}\sqrt{26/25} - 20\log_{10}\sqrt{2} - 20\log_{10}\sqrt{101/100}$$
$$\approx 40.6 \text{ dB}$$

The next corner frequency is given by $\omega_{c2} = 5$ where we find that

$$20\log_{10}|150| + 20\log_{10}|1 + j| - 20\log_{10}|1 + j5| - 20\log_{10}|1 + j\omega/2|$$
$$= 20\log_{10}|150| + 20\log_{10}\sqrt{2} - 20\log_{10}\sqrt{26} - 20\log_{10}\sqrt{5/4} \approx 30.4 \text{ dB}$$

Finally, the last corner frequency is at $\omega_{c3} = 10$. In this case we find

$$20\log_{10}|150| + 20\log_{10}|1 + j2| - 20\log_{10}|1 + j10| - 20\log_{10}|1 + j|$$
$$= 20\log_{10}|150| + 20\log_{10}\sqrt{5} - 20\log_{10}\sqrt{101} - 20\log_{10}\sqrt{2} \approx 27.5 \text{ dB}$$

Returning to the original expression, we had

$$|H(\omega)|_{\text{dB}} = 20\log_{10}|150| + 20\log_{10}|1 + j\omega/5| - 20\log_{10}|1 + j\omega| \\ - 20\log_{10}|1 + j\omega/10|$$

The first term is $20\log_{10}|150| \approx 44$ dB, which adds a constant or piston term to the plot. To generate the Bode plot, we add in each term at the appropriate corner frequency. Since the first corner frequency occurs at $\omega_{c1} = 1$, up to that point we have the constant term

$$|H(\omega)|_{\text{dB}} = 20\log_{10}|150| \quad 0 < \omega \leq 1$$

Next, between $\omega_{c1} = 1$ and $\omega_{c2} = 5$, we add the second term whose corner frequency is $\omega_{c2} = 5$, giving

$$|H(\omega)|_{\text{dB}} = 20\log_{10}|150| - 20\log_{10}|1 + j\omega| \quad 1 < \omega \leq 5$$

The next corner frequency occurs at $\omega_{c3} = 10$, up until this point we add in the next term

$$|H(\omega)|_{\text{dB}} = 20\log_{10}|150| + 20\log_{10}|1 + j\omega/5| - 20\log_{10}|1 + j\omega| \\ 5 < \omega \leq 10$$

The last part of the plot is for $\omega > 10$, where we add in the last term to obtain

$$|H(\omega)|_{\text{dB}} = 20\log_{10}|150| + 20\log_{10}|1 + j\omega/5| - 20\log_{10}|1 + j\omega| \\ - 20\log_{10}|1 + j\omega/10|$$

The plot is shown below in Fig. 15-5.

Filters

Characterizing the plot of $|H(j\omega)|$ for a filter is so important that we revisit it here. Let's consider some examples.

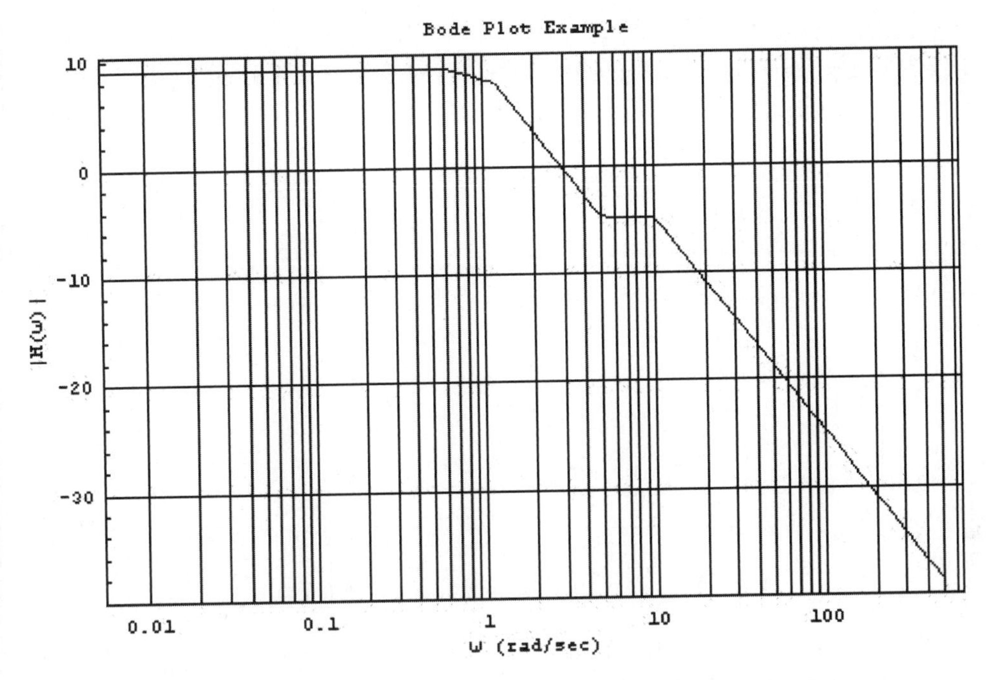

Fig. 15-5 A Bode plot of the transfer function of Example 15-4.

EXAMPLE 15-5

A filter has a transfer function given by

$$H(s) = \frac{1}{s+4}$$

Plot its magnitude versus frequency. What type of filter does this represent?

SOLUTION

First, looking at the function notice that the function has a simple pole at $s = -4$ since

$$\lim_{s \to -4} \frac{1}{s+4} = \infty$$

Also notice that

$$\lim_{s \to \infty} \frac{1}{s+4} = 0$$

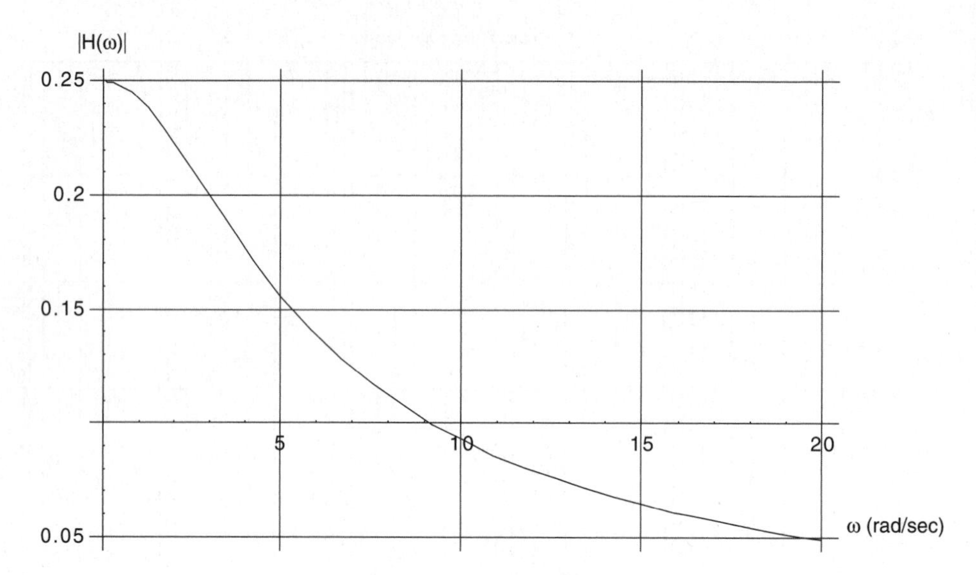

Fig. 15-6 A plot of the magnitude of $H(s) = \frac{1}{s+4}$.

So we say that this transfer function has a simple zero at $s \to \infty$. We can see that this is not a band-pass filter because as $s \to 0$, $H(s) \to 1/4$. Now let $s \to j\omega$. Then

$$H(j\omega) = \frac{1}{4 + j\omega} = \frac{1}{4 + j\omega}\left(\frac{4 - j\omega}{4 - j\omega}\right) = \frac{4 - j\omega}{\omega^2 + 16}$$

Hence

$$|H(j\omega)| = \sqrt{\frac{16}{(\omega^2 + 16)^2} + \frac{\omega^2}{(\omega^2 + 16)^2}} = \frac{1}{\sqrt{\omega^2 + 16}}$$

Let's plot this. The plot is shown in Fig. 15-6, from which it's clear that this circuit will function as a low-pass filter.

Butterworth Filters

A *Butterworth* low-pass filter has a transfer function of the form

$$|H(\omega)| = \frac{1}{\sqrt{1 + \left(\frac{\omega}{\omega_c}\right)^{2n}}} \tag{15.10}$$

The number of inductors and capacitors in the circuit used to construct the filter is given by n. We call $n$the *order* of the Butterworth filter. By increasing n, we can get closer to an ideal low-pass filter with a sharp cutoff. This is because

$$\lim_{n\to\infty}\left(\frac{\omega}{\omega_2}\right)^{2n} = \infty \tag{15.11}$$

when $\omega > \omega_c$. Then

$$\lim_{n\to\infty}|H(\omega)| = \lim_{n\to\infty}\frac{1}{\sqrt{1+\left(\frac{\omega}{\omega_c}\right)^{2n}}} = \frac{1}{\sqrt{1+\infty}} \to 0$$

Hence, high frequencies are not passed through the filter. On the other hand, when $\omega < \omega_c$

$$\lim_{n\to\infty}\left(\frac{\omega}{\omega_2}\right)^{2n} = 0$$

Therefore

$$\lim_{n\to\infty}|H(\omega)| = \lim_{n\to\infty}\frac{1}{\sqrt{1+\left(\frac{\omega}{\omega_c}\right)^{2n}}} = \frac{1}{\sqrt{1+0}} \to 1$$

So for frequencies below the cutoff, the transmission is "perfect," as if the filter was described by a transfer function given by a unit step with cutoff at ω_c. A Butterworth filter of order 1 has a transfer function given by

$$H(s) = \frac{1}{s+1}$$

Since

$$|H(j\omega)| = \frac{1}{\sqrt{1+\omega^2}}$$

Here we are implicitly setting the cutoff frequency to 1.

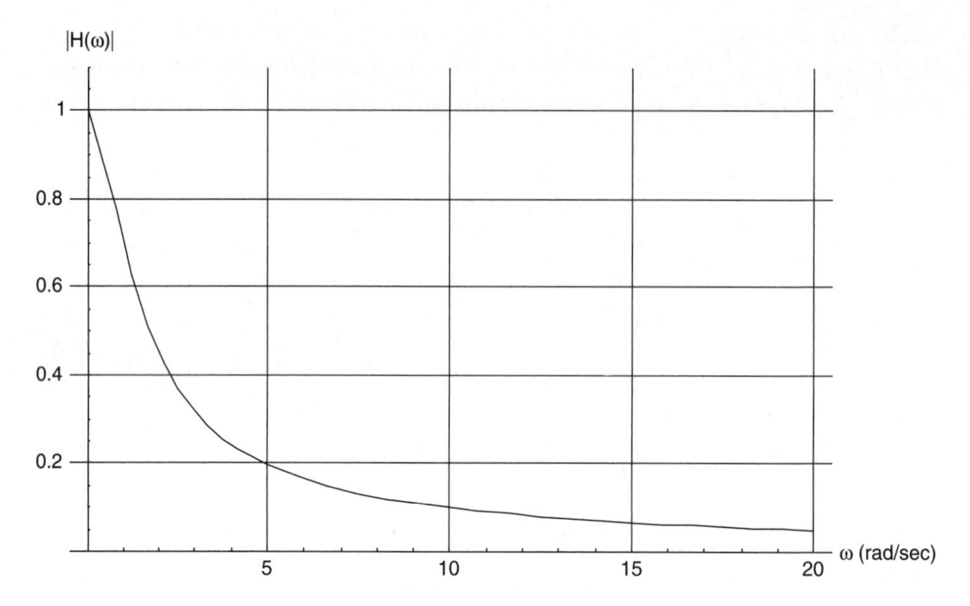

Fig. 15-7 A plot of a first-order Butterworth filter.

Let's verify the improved performance of a Butterworth filter as n gets larger. We start with the first-order filter, showing a plot in Fig. 15-7.

This filter is far from the ideal case; the cutoff drops gradually rather than sharply. Now let's let $n = 4$. In this case the transfer function is given by

$$|H(\omega)| = \frac{1}{\sqrt{1 + \omega^8}}$$

The characteristics of the filter are significantly better already. The response of the filter drops quickly over a small frequency range, and in practice it might be enough for many purposes. A plot of a fourth-order Butterworth filter is shown in Fig. 15-8.

As n gets even larger, the behavior of the filter quickly approaches the ideal case. In Fig. 15-9, we show a plot for a twentieth-order Butterworth filter. This is an essentially ideal low-pass filter, with a transfer function that rapidly drops to zero at the cutoff frequency $\omega_c = 1$.

It is often preferred to have a low-pass filter with a cutoff of a desired sharpness. That is, we may specify how rapidly the magnitude of the transfer function drops off. This can be specified by choosing the order n of the Butterworth filter. To determine the order of a Butterworth filter we begin by considering the limit of the transfer function at high frequency. Looking at (15.10) and letting $\omega_c = 1$ without loss of generality, notice that as ω gets large we can ignore the 1 in the

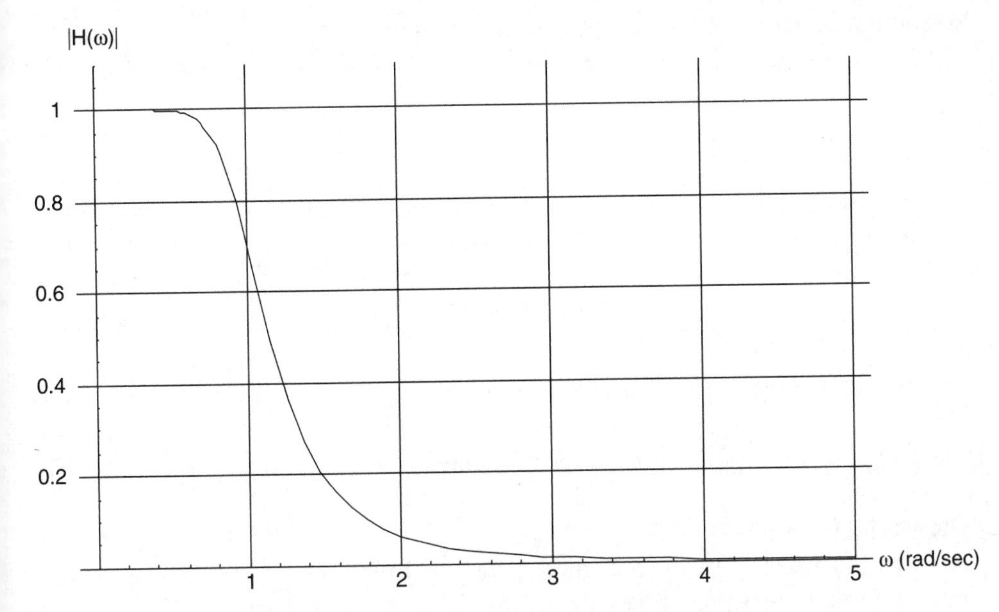

Fig. 15-8 A plot of a fourth-order Butterworth filter.

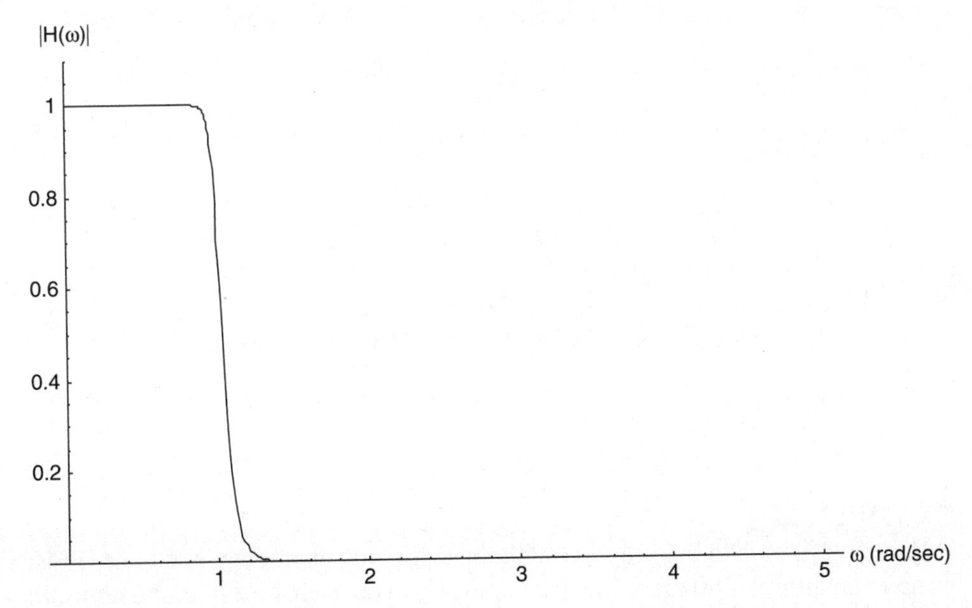

Fig. 15-9 A twentieth-order Butterworth filter is essentially an ideal low-pass filter.

denominator. Considering the gain function, then we have

$$\lim_{\omega \to \infty} 20 \log_{10} \frac{1}{\sqrt{\omega^{2n}}} = -20n \log_{10} \omega$$

Remember that a decade is defined by using (15.4), so this tells us that a Butterworth filter has a dropoff or attenuation of

$$20n \text{ dB/decade} \tag{15.12}$$

Or the attenuation can be described as

$$6n \text{ dB/octave} \tag{15.13}$$

EXAMPLE 15-6

A low-pass filter is to be designed with the following characteristic. There must be an attenuation of 390 dB at the frequency given by $\omega = 20\omega_c$. Find the required order for the circuit and write down the magnitude of the transfer function.

SOLUTION

Since $\omega = 20\omega_c$ this tells us that we are two decades past the critical frequency. With an attenuation of $20n$ dB/decade, the order of our circuit must satisfy

$$40n \geq 390$$

That is

$$n \geq 9.75$$

The order of a Butterworth filter is an integer, so we choose the smallest integer satisfying this inequality, $n = 10$. The transfer function is given by

$$|H(\omega)| = \frac{1}{\sqrt{1 + \omega^{20}}}$$

EXAMPLE 15-7

A low-pass filter is to be designed with the following characteristic. There must be an attenuation of 80 dB at the frequency given by $\omega = 6\omega_c$. Find the required order for the circuit and write down the magnitude of the transfer function.

SOLUTION

Notice that $6\omega_c$ is three *octaves* beyond the cutoff frequency. With an attenuation of 6 dB per octave, we have

$$18n \geq 80$$

Hence

$$n \geq 4.4$$

Therefore we must choose $n = 5$. The magnitude of the transfer function is

$$|H(\omega)| = \frac{1}{\sqrt{1 + \omega^{10}}}$$

Quiz

1. Plot $|H(\omega)|_{\text{dB}}$ for $H(\omega) = 1 + j\frac{\omega}{30}$
2. What are the corner frequencies for

$$H(\omega) = 1000\frac{1 + j\omega}{(1000 + 110j\omega - \omega^2)}$$

3. Construct a Bode plot for the transfer function of Problem 2.
4. A low-pass filter is to be designed with the following characteristic. There must be an attenuation of 60 dB at the frequency given by $\omega = 4\omega_c$. Find the required order for the circuit and write down the magnitude of the transfer function.

Final Exam

1. You establish an observation point in a wire and find that $q(t) = (8t^3 - 2t)$ nC. Find the current flowing past your observation point.
2. If $q(t) = 5 \sin 4t$ mC, what is the corresponding current?
3. If the current is $i(t) = 5 \sin 4t$, where current is given in amps, how much charge flows by between 0 and 1 s?
4. At a certain point P in a wire, 135 C of positive charge flow to the right while 75 C of negative charge flow to the left. What is the current flowing in the wire?
5. A charge $q = 2$ C passes through a potential difference of 8 V. How much energy does the charge acquire?
6. If the voltage in a circuit is given by $v(t) = 3 \cos 126t$, what is are the amplitude and cycles per second?
7. In some circuit element the power is 8 W and the voltage is 1 V. How much current flows?
8. Find the power in each element shown in Fig. FE-1.

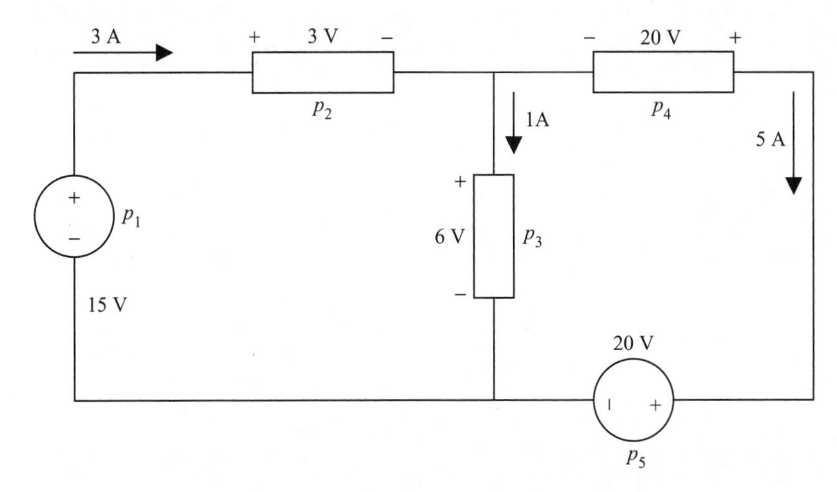

Fig. FE-1 Circuit diagram for Problem 8.

9. How does conservation of energy manifest itself in a circuit?
10. Find the missing power in Fig. FE-2.
11. Consider the node shown in Fig. FE-3. Find the current i_3 if $i_1 = -1$ A and $i_2 = 3$ A.
12. Consider the circuit shown in Fig. FE-4. Find the unknown voltage.
13. Find the unknown voltages in Fig. FE-5.
14. It is known that the voltage across a resistor is 10 V, while 2 A of current flows through the resistor. What is the resistance?
15. In a circuit, 3 A of current flows through a 8 Ω resistor. What is the voltage? What is the conductance of the resistor?

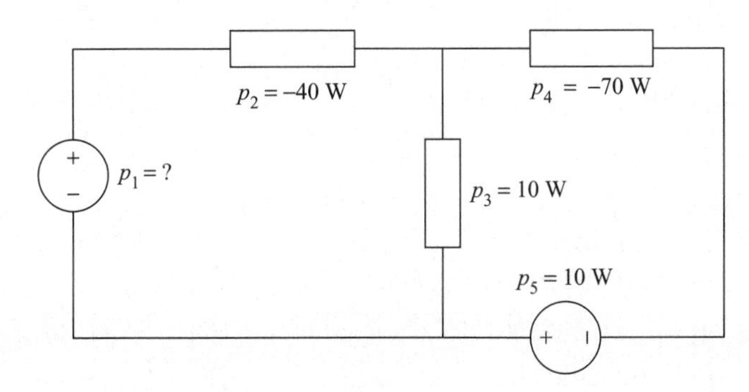

Fig. FE-2 Circuit diagram for Problem 10.

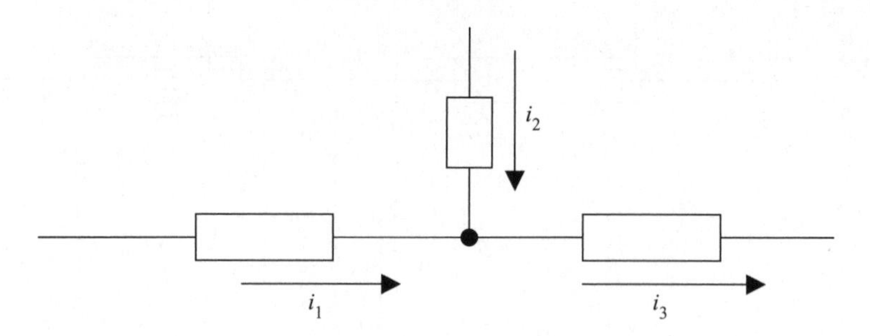

Fig. FE-3 Circuit diagram for Problem 11.

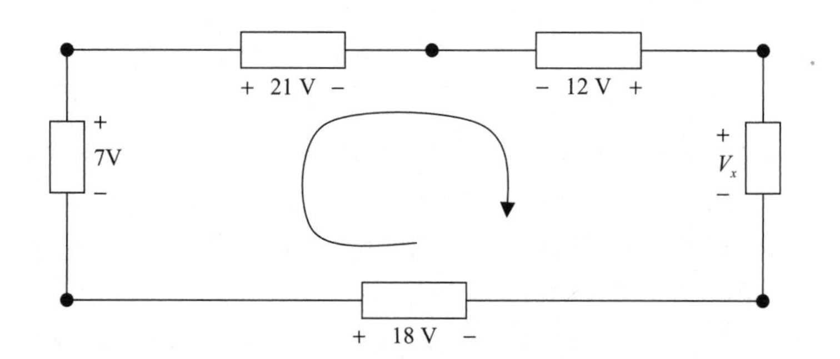

Fig. FE-4 Circuit loop for Problem 12.

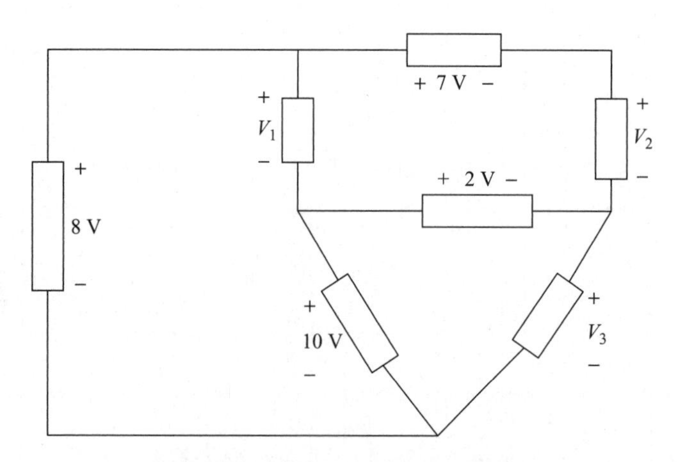

Fig. FE-5 Circuit for Problem 13.

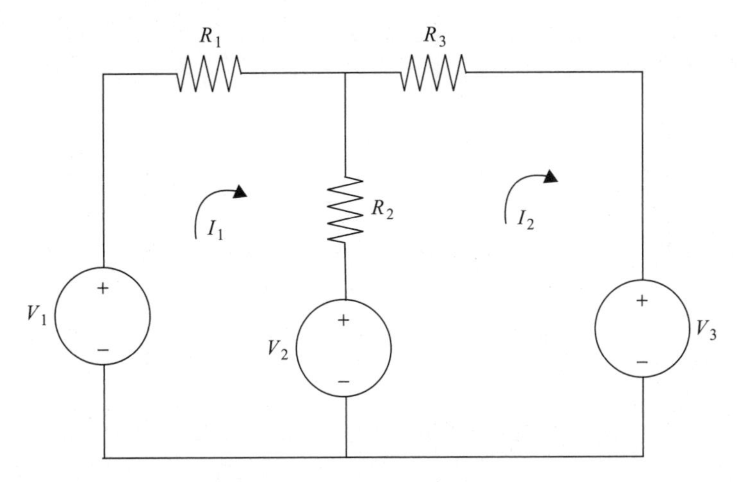

Fig. FE-6 The circuit Problem 16.

16. Consider the circuit shown in Fig. FE-6. Find the unknown currents for the circuit in Fig. 2-17. Suppose that $R_1 = 2\ \Omega,\ R_2 = 1\ \Omega$, and $R_3 = 3\ \Omega$ and $V_1 = 10$ V, $V_2 = 3$ V, and $V_3 = 6$ V.
17. A 17 Ω resistor is in series with a 12 Ω resistor. What is the equivalent resistance?
18. Two 10 Ω resistors are in parallel. What is the equivalent resistance?
19. Find the equivalent resistance for the circuit shown in Fig. FE-7.
20. How is the Norton current related to the Thevenin equivalent voltage?
21. Apply the Karni method to the circuit shown in Fig. FE-8. Find the equation for v_o and use it to write down the Thevenin equivalent voltage and resistance as seen by the load resistor R_L.

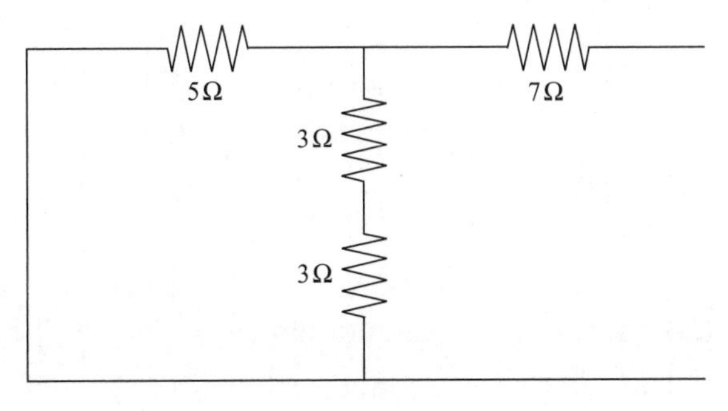

Fig. FE-7 Circuit for Problem 19.

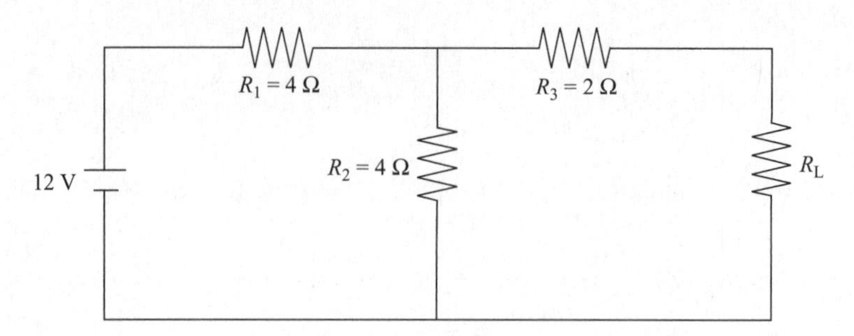

Fig. FE-8 In Problem 21, apply the Karni method to this circuit.

22. When applying superposition to a circuit containing voltage sources, they are
    A. Replaced by open circuits.
    B. Set to zero.

23. When applying superposition to a circuit containing dependent voltage sources, they are
    A. Left alone.
    B. Set to zero.
    C. Replaced by open circuits.
    D. Replaced by short circuits.

24. When solving a circuit using superposition
    A. Power can be calculated individually with each source set to zero, then summed.
    B. Power can be calculated due to each source individually, then the total power is found from the product of the individual powers.
    C. The superposition theorem cannot be applied to power calculations.

25. Refer to Fig. 4-1. Use superposition to find the current flowing in the 5 Ω resistor. Then determine what power this resistor absorbs.
26. Refer to the circuit shown in Fig. 4-x. Use superposition to find the current flowing through the 200 Ω resistor.
27. Three resistors $R = 3$ are connected in a delta configuration. What is R for the equivalent Y configuration?
28. Three resistors $R = 3$ are connected in a Y configuration. What is R for the equivalent delta configuration?

29. In a Wheatstone bridge with $R_1 = 3$, $R_3 = 8$ it is found that balance is achieved when $R_2 = 7$. What is the value of the unknown resistance?
30. A 0.2 F capacitor is in an RC circuit with a 100 Ω resistor. What is the time constant?
31. A 0.2 F capacitor is in an RC circuit with a 100 Ω resistor. How long does it take for all the voltages and currents to decay to zero?
32. Find the inductance of a 500 turn coil linked by a 2×10^{-5} Wb flux when a 10 mA current flows through it.
33. Find the coefficient of coupling between a 0.2 H inductor and a 0.3 H inductor when $M = 0.18$.
34. A current increases uniformly from 1 to 5 A in a coil, over a period of 3 s. This induces a voltage of 5 V across the coil. What is the inductance of the coil?
35. A 0.5 F capacitor is in series with a 10 V dc voltage source and a 1 Ω resistor. Find the voltage across the capacitor as a function of time if the initial voltage is zero.
36. Consider an RL circuit with $R = 10\ \Omega$, $L = 4$ H in series with a voltage source with $v(t) = 4$. Find the total solution. Assume the initial current is zero.
37. Consider an RL circuit with $R = 10\ \Omega$, $L = 4$ H in series with a voltage source with $v(t) = 4t$. Find the total solution. Assume the initial current is zero.
38. Consider an RL circuit with $R = 10\ \Omega$, $L = 4$ H in series with a voltage source with $v(t) = 4t$. Find the total solution. Assume the initial current is $i(t) = -1$ A.
39. Consider an RL circuit with $R = 10\ \Omega$, $L = 4$ H in series with a voltage source with $v(t) = 4\cos t$. Find the total solution. Assume the initial current is $i(t) = 0$ A.
40. Consider an RL circuit with $R = 10\ \Omega$, $L = 5$ H in series with a voltage source with $v(t) = 5t^2$. Find the total solution. Assume the initial current is $i(t) = 0$ A.
41. A load has a voltage $\mathbf{V} = 40\angle 10^\circ$ and current $\mathbf{I} = 10\angle 0^\circ$. Find the impedance and determine a series circuit that will model the load. Is the circuit inductive or capacitive? Assume that $\omega = 100$ rad/s.
42. A circuit has a given transfer function $\mathbf{H}$. What is the condition for resonance?

43. What is the admittance in terms of conductance and susceptance?
44. In terms of phasors, how is the transfer function defined?
45. A circuit has a transfer impedance. Describe the relation between input and output.
46. What is the most important characteristic of the transfer function at resonance?
47. A transfer function is given by $\mathbf{H}(\omega) = \dfrac{j\omega C}{(1 - \omega^2 LC + j\omega RC)}$. What is the resonant frequency?
48. Define a high-pass filter.
49. If a circuit is critically damped, how is the damping factor related to the resonant frequency?
50. Write $z = 5\sqrt{2}(1 + i)$ in polar form.
51. Let $v_0(t) = A\cos\omega t$, $v_1(t) = A\cos(\omega t + 180°)$. Are the waveforms in phase?
52. If the current flowing in a resistor is $i(t) = I\sin\omega t$, determine the average power.
53. Find the instantaneous power in a capacitor with a voltage $v(t) = V\sin(\omega t + \phi)$ across it.
54. A current $i(t) = I\sin(\omega t + \phi)$ flows through an inductor. What is the instantaneous power?
55. What is the reactance of an inductor if $i(t) = I\sin(\omega t + \phi)$?
56. A circuit consists of a voltage source, inductor L and capacitor C arranged in series. Derive the differential equation that can be solved to obtain the zero-input response of the circuit.
57. Continue with the circuit in Problem 56. What is the natural frequency of the circuit?
58. What is admittance of the circuit in Problem 56?
59. What type of filter is described by the differential equation $V_0\cos\omega t = v_c + RC\dfrac{dv_c}{dt}$?
60. The instantaneous power in a circuit is $p = 20 + 10\cos(377t + 40°)$. Find the maximum, minimum, and average power.
61. A load has a voltage $v = 300\cos(20t + 30°)$ applied and draws a current $i = 15\cos(20t - 25°)$. What is the power factor?

62. A load has a voltage $v = 170 \sin(377t)$ and draws a current $i = 20 \sin(377t - 10°)$. What is the power factor? Is it leading or lagging?
63. A load has a voltage $v = 120\sqrt{2} \sin(377t + 10°)$ and draws a current $i = 12\sqrt{2} \sin(377t + 30°)$. What is the power factor? Is it leading or lagging?
64. What is the power factor if a circuit absorbs 600 W for a 220 V input and a 20 A current, where the voltage and current are given as effective values?
65. An effective voltage of 110 V is applied to a load. The impedance of the load is $Z = 10\angle 20°$. What is the absorbed power?
66. Determine the impedance of a circuit constructed with a resistor and capacitor in parallel, if it is connected to a household outlet at 120 V, 60 Hz. The circuit absorbs 60 W and p.f. = 0.8 lagging.
67. What is the reactive power if $V = 120$ V, $I = 12$ A, and p.f. = 0.8 lagging?
68. What is the magnitude of the apparent power if $V = 120$ V and $I = 12$ A?
69. What is the average power for a purely inductive load and a sinusoidal voltage source?
70. Find the Laplace transform of the unit impulse or Dirac delta function $\delta(t)$.
71. Find the Laplace transform of $f(t) = \sin \omega t$.
72. What is the Laplace transform of $f(t) = e^{-5t}u(t)$?
73. Is $f(t) = \sin 3t$ of exponential order?
74. Find the Laplace transform of $u(t - a)$.
75. Find the inverse Laplace transform of $F(s) = \dfrac{1}{s^2(s+1)}$.
76. Find the inverse Laplace transform of $F(s) = \dfrac{s-2}{(s-4)(s^2+16)}$.
77. What are the poles and zeros of $F(s) = \dfrac{s}{(s-2)(s+1)}$?
78. The unit impulse response of a circuit is $h(t) = t$. Is the circuit stable?
79. The unit impulse response of a circuit is $h(t) = te^{-3t}$. Is the circuit stable or unstable?

80. The unit impulse response of a circuit is $h(t) = t^4 e^{-3t}$. Is the circuit impulse response stable?

81. The transfer function of a circuit is $H(s) = \dfrac{1}{s-1}$. Is the circuit impulse response stable?

82. The transfer function of a circuit is $H(s) = \dfrac{1}{s+2}$. Is the circuit impulse response stable?

83. The transfer function of a circuit is $H(s) = \dfrac{16}{s(s^2+8s+16)}$. Is the circuit impulse response stable?

84. The transfer function of a circuit is $H(s) = \dfrac{16}{s(s^2+8s+16)}$. Is the circuit impulse response stable if we require that $\lim_{t\to\infty} |h(t)| = 0$?

85. The transfer function of a circuit is $H(s) = \dfrac{1}{(s+4)^2}$. What is the impulse response of the circuit? Is the circuit stable?

86. The transfer function of a circuit is $H(s) = \dfrac{1}{(s-1)(s+2)}$. Is the circuit stable?

87. The transfer function of a circuit is $H(s) = \dfrac{1}{s^2-36}$. Is the circuit stable?

88. The transfer function of a circuit is $H(s) = \dfrac{1}{s^2+36}$. Is the circuit stable?

89. Define BIBO stability.

90. An RLC circuit is excited with a sinusoidal source. When is it not BIBO stable?

91. What is the magnitude of the frequency response in decibels?

92. Two frequencies are an octave apart. How are they related?

93. What is the low-frequency asymptote of $H(s) = \dfrac{2s+6}{s^2-s-12}$?

94. How does the phase angle of $H(s) = \dfrac{2s+6}{s^2-s-12}$ behave at high frequencies?

95. Describe the vertical axis of a Bode plot.

96. Where is the cutoff frequency located in a Bode plot?
97. Suppose that $H(\omega) = 300\frac{5 + j\omega}{-\omega^2 + j11\omega_- + 10}$. Where are the corner frequencies?
98. The transfer function of a filter is given by $H(s) = \frac{1}{s + 5}$. What kind of filter is this?
99. Suppose that $|H(\omega)| = \frac{1}{\sqrt{1 + \omega^8}}$. Is this a Butterworth filter? Of what order?
100. What is the attenuation per octave of an nth-order Butterworth filter?

Quiz and Exam Solutions

Chapter 1

1. $i = 2\text{ A}$
2. $i(t) = -10e^{-2t}(2\cos 5t + 5\sin 5t)$

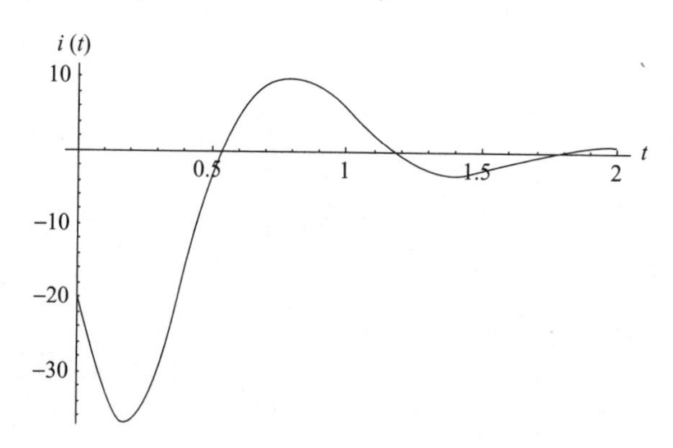

Fig. Q-1 Plot of $i(t) = -10e^{-2t}(2\cos 5t + 5\sin 5t)$.

3. 2.25 C

4. 28 A
5. 56 J
6. 120 V, 100 cps
7. 2 A
8. −15 W, 3 W, 12 W, −100 W, 100 W
9. $\sum p_i = 0$
10. 40 W

Chapter 2

1. $i_1 = -3$ A
2. $V_x = -14$ V
3. $V_1 = -14$ V, $V_2 = 4$ V, $V_3 = 13$ V
4. 5 Ω
5. $V = 100$ V, $G = 0.2$ S
6. 28 W dissipated
7. 408 W
8. $i(t) = -\sin 10t$
9. $v_1(t) = \dfrac{R_1}{R_1 + R_2} v_s(t)$
10. $I_1 = 0.66$ A, $I_2 = 0.55$ A, $I_3 = 0.97$ A

Chapter 3

1. 11 Ω
2. 2.4 Ω
3. 10.5 Ω
4. $R_{TH} = 6.6 \quad \Omega, \quad V_{TH} = -15.6$ V
5. $i(t) = -2.4 + 0.3e^{-t} \quad [A]$

Chapter 4

1. 92 V
2. No, power is nonlinear.

3. $V_M = \dfrac{G_1V_1 + G_2V_2 + \cdots + G_nV_n}{G_1 + G_2 + \cdots + G_n}$
4. $R_M = \dfrac{1}{G_1 + G_2 + \cdots + G_n}$
5. $V_M = 2\text{ V},\ R_M = 6/11\ \Omega$

Chapter 5

1. $R_A = R_B = R_C = 4\ \Omega$
2. $R = 3\ \Omega$
3. $R_4 = 12\ \Omega$

Chapter 6

1. 0.01 ms
2. 0.05 ms
3. 0.44 H
4. $i(t) = 3e^{-5t}(e^{4t} - 1)$
5. $\dfrac{1}{52}(\cos 10t + 5\sin 10t - e^{-2t})$
6. $i(t) = 2t - 1 + 3e^{-2t}$
7. $i = \dfrac{1}{15}e^{-t/4}\left[15\cos\left(\dfrac{\sqrt{15}t}{4}\right) + \sqrt{15}\sin\left(\dfrac{\sqrt{15}t}{4}\right)\right]$

Chapter 7

1. $z = 3e^{j5\pi/6}$
2. Ampltude = 12. The waves are not in phase. $i_2(t)$ leads $i_1(t)$ by 10°.
3. $p = \dfrac{V^2}{2R}$
4. $Z = V/I$
5. $i(t) = 7.84\sin(4t - 13.7°)$
6. $\mathbf{V}_{\text{TH}} = \mathbf{Z}_{\text{TH}}\mathbf{I}_0 + \mathbf{V}_0$

Chapter 8

1. Capacitive, $C = 2.94$ mF
2. Solution obtains $\omega = \pm\frac{j}{\sqrt{10}}$; however, frequencies must be real and positive. Hence the circuit cannot have the voltage and current in phase.
3. $\omega = \sqrt{\frac{C - L/R_B^2}{LC^2}}$
4. Yes, the transfer function is a low-pass filter.
5. $\Delta\omega = 25$, $Q = 25.13$

Chapter 9

1. -200 V
2. 21
3. 210 V
4. -7 V

Chapter 10

1. 0
2. $p(t) = \frac{V_0 I_0}{2}\cos(2\omega t + 90°)$
3. $C = \frac{1}{\omega^2}L$
4. 1.6 mF
5. $I_1 = 1.14$ A, $I_2 = 0.91$ A

Chapter 11

1. $8\angle 0°$
2. $16(1 + j)$
3. $i_1(t) = 8.8\cos(100t - 45°)$

Chapter 12

1. $V_A = 10\angle -140°$, negative
2. b

Chapter 13

1. $F(s) = 1/s$
2. $F(s) = \dfrac{s}{s^2 + \omega^2},\ s > 0$
3. $F(s) = \dfrac{2}{(s-1)^2 + 4}$
4. $F(s) = \dfrac{72}{s^5} + \dfrac{5}{s},\ G(s) = \dfrac{10 - 3s}{s^2 + 4}$
5. No, because we cannot find any a such that $te^{t^2}e^{-at} \to 0$ as $t \to \infty$, because e^{t^2} blows up faster than e^{-at}.
6. $f(t) = 2\cos 5t - \dfrac{2}{5}\sin 5t$
7. $f(t) = \dfrac{1}{15}e^{-3t}(7e^{5t} + 20e^{2t} - 27)$
8. $f(t) = \dfrac{1}{5}(3e^{2t} - 3\cos t - \sin t)$
9. Zero state: $4(2 - 2e^{-t/2})$, zero input: $3e^{-t/2}$.
10. $i(t) = \dfrac{1}{25}(8\cos 2t + 6\sin 2t + 5te^{-t} - 8e^{-t})$
11. $f(t) = e^{-t} + \cos t - \sin t$
12. $h(t) = \dfrac{1}{RC}e^{-t/RC},\ H(s) = \dfrac{1}{1 + sRC}$
13. $r(t) = -\dfrac{\omega}{RC + \omega^2}\sin(\omega t)u(t) - \dfrac{1}{RC + \omega^2}\left(\dfrac{1}{RC}\right)e^{-t/RC}u(t),$

    $$R(s) = -\left(\frac{\omega}{RC + \omega^2}\right)\left(\frac{\omega}{s^2 + \omega^2}\right) - \left(\frac{1}{RC + \omega^2}\right)\left(\frac{1}{1 + sRC}\right)$$
14. Poles : $s = -1, \pm 2i$

Chapter 14

1. Unstable
2. $h(t) = \cos 4t$, stable
3. $h(t) = e^{6t}$, unstable
4. $h(t) = tu(t)$, unstable
5. $h(t) = e^{-2t}\sin 6t$, stable
6. Unstable, $\lim_{t\to\infty} v_c(t) = \infty$

Chapter 15

1. The plot is given by

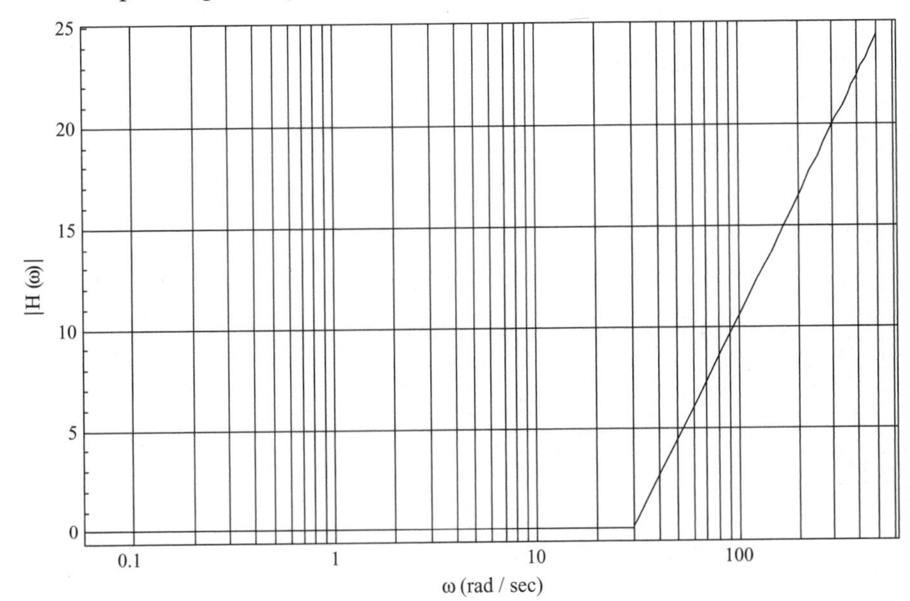

2. $\omega_c = 1{,}10{,}100$
3. The plot is given by

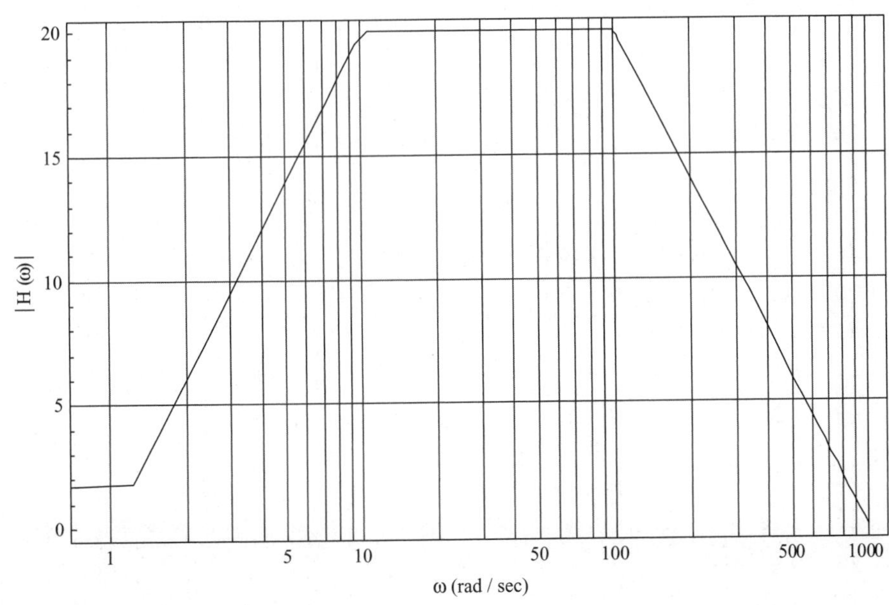

4. $|H(\omega)| = \dfrac{1}{\sqrt{1+\omega^{10}}}$

Final Exam

1. $i = 24t^2 - 2\text{nA}$
2. $i(t) = 20 \sin 4t$ mA.
3. 2.1 C
4. 210 A
5. 16 J
6. 3 V, 40 cps
7. 8 A
8. −15 W, 9 W, 6 W, −100 W, 100 W
9. $\sum p_i = 0$
10. 90 W
11. 2 A
12. 16 V
13. $V_1 = -2$ V, $V_2 = -7$ V, $V_3 = 8$ V
14. 5 Ω
15. $V = 24$ V, $G = 0.13$ S
16. $I_1 = 2.3$ A, $I_2 = -0.2$ A
17. 29 Ω
18. 5 Ω
19. 9.7 Ω
20. $I_N = \frac{V_{TH}}{R_{TH}}$
21. $v_o = 6 + 4I_o, \quad V_{TH} = 6, \; R_{TH} = 4$
22. B
23. A
24. C
25. 2.7 W
26. −0.16 A
27. 1 Ω
28. 9 Ω
29. $R_4 = 19$ Ω

30. 20 s
31. 1 min 40 s
32. 1 H
33. 0.72
34. 3.75 H
35. $v_c(t) = 10(e^{-2t} - 1)$
36. $i(t) = \frac{2}{5}(1 - e^{-5t/2})$
37. $i(t) = \frac{8}{25}(5t - 2 + 2e^{-5t/2})$
38. $i(t) = \frac{1}{25}(10t - 4 - 21e^{-5t/2})$
39. $\frac{2}{29}(2 \sin t + 5 \cos t - 5e^{-5t/2})$
40. $\frac{1}{4}(1 - 2t + 2t^2 - e^{-2t})$
41. Inductive, L = 7 mH
42. When the phase angle of the transfer function vanishes.
43. $Y = G + jB$
44. $\mathbf{R} = \mathbf{HE}$
45. $\mathbf{I} = \mathbf{YV}$
46. $\theta_H = 0$
47. $\omega = \frac{1}{\sqrt{LC}}$
48. Does not allow frequencies where $\omega < \omega_c$ to pass through.
49. $\varsigma = \omega_0$
50. $z = 10e^{j\pi/4}$
51. No v_0 lags v_1
52. $p = \frac{I^2 R}{2}$
53. $p = \frac{VI}{2} \sin(2\omega t + 2\phi)$
54. $(V_{\text{eff}})(I_{\text{eff}}) \sin(2\omega t + 2\phi)$
55. $X_L = \omega L$
56. $LC\frac{d^2 v_C}{dt^2} + v_C = 0$
57. $\omega_0 = \frac{1}{\sqrt{LC}}$
58. $\mathbf{Y} = \frac{1}{\mathbf{Z}}$
59. This circuit can be a high-pass filter.

60. Max = 30 W, min = 10 W, avg = 20 W
61. p.f. = 0.574
62. p.f. = 0.98 lagging
63. p.f. = 0.93 leading
64. p.f. = 0.25
65. Absorbed power is 1137 W.
66. $R = 240$, $C = 3.1$ mF, $Z = R + j\omega C$
67. $Q = 864$
68. 1440
69. 0
70. $F(s) = 1$
71. $F(s) = \dfrac{\omega}{s^2 + \omega^2}$
72. $F(s) = \dfrac{1}{s - 5}$
73. Yes, for any $a > 0$
74. $F(s) = \dfrac{e^{-as}}{s}$
75. $f(t) = t - 1 + e^{-t}$
76. $f(t) = \dfrac{1}{16}(e^{4t} - \cos 4t + 3 \sin 4t)$
77. Zeros $s = 0$, poles $s = 2, s = -1$
78. Unstable
79. Stable
80. Stable
81. Unstable
82. Stable
83. Yes
84. No
85. $h(t) = te^{-4t}$, stable
86. Unstable
87. No

88. The circuit is impulse response stable, not BIBO stable if there is an input resonance.
89. Bounded input, bounded output stability. The circuit has a bounded response given a bounded input.
90. If the input has a frequency that matches the natural frequency of the circuit, there is resonance. Even though there is a bounded input (a sinusoidal function) the output will blow up.
91. $|H(\omega)|_{\text{dB}} = 20\ \log_{10}|H(\omega)|$
92. $\omega_A = 2\omega_B$
93. $s = -\frac{1}{2}$
94. $\theta = \lim_{\omega\to\infty} \tan^{-1}\frac{\omega}{4} = 90^\circ$
95. The vertical axis is the frequency response in decibels.
96. The intersection with the 0 dB axis.
97. 1,5,10
98. Low-pass
99. Yes, $n = 4$
100. $6n$ dB/octave

References

Horowitz, Paul and Hill, Winfield, *The Art of Electronics,* 2nd ed., Cambridge University Press, Cambridge, U.K., 1989.

Hsu, Hwei, *Schaum's Outlines: Signals and Systems,* McGraw-Hill, New York, 1995.

Karni, Shlomo, *Applied Circuit Analysis,* John Wiley & Sons, New York, 1988.

O'Malley, John, *Schaum's Outlines: Basic Circuit Analysis,* 2nd ed., McGraw-Hill, New York, 1992.

INDEX

Note: Information presented in figures and tables is denoted by t and f, respectively.

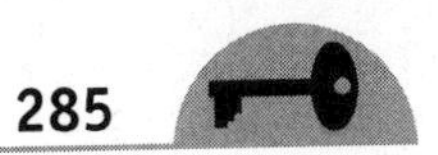